Cambridge International AS & A Level Mathematics

Pure Mathematics 2 & 3

STUDENT'S BOOK

Tom Andrews, Helen Ball,
Michael Kent, Chris Pearce
Series Editor: Dr Adam Boddison

William Collins' dream of knowledge for all began with the publication of his first book in 1819.

A self-educated mill worker, he not only enriched millions of lives, but also founded a flourishing publishing house. Today, staying true to this spirit, Collins books are packed with inspiration, innovation and practical expertise. They place you at the centre of a world of possibility and give you exactly what you need to explore it.

Collins. Freedom to teach.

Published by Collins
An imprint of HarperCollins*Publishers*
The News Building
1 London Bridge Street
London
SE1 9GF

HarperCollinsPublishers
Macken House, 39/40 Mayor Street Upper,
Dublin 1,
D01 C9W8,
Ireland

MIX
Paper | Supporting
responsible forestry
FSC™ C007454
www.fsc.org

This book is produced from independently certified FSC™ paper to ensure responsible forest management.

For more information visit:
www.harpercollins.co.uk/green

Browse the complete Collins catalogue at
www.collins.co.uk

British Library Cataloguing in Publication Data

A catalogue record for this publication is available from the British Library.

Commissioning editor: Jennifer Hall
In-house editor: Lara McMurray
Authors: Tom Andrews/Helen Ball/Michael Kent/Chris Pearce
Series editor: Dr Adam Boddison
Development editor: Tim Major
Project manager: Emily Hooton
Copyeditor: Jan Schubert
Reviewer: Adele Searle
Proofreaders: Julie Bond/Joan Miller
Answer checkers: Deborah Dobson/Steven Matchett
Cover designer: Gordon MacGilp
Cover illustrator: Maria Herbert-Liew
Typesetter: Jouve India Private Ltd
Illustrators: Jouve India Private Ltd/Ken Vail Graphic Design
Production controller: Sarah Burke
Printed and bound by Ashford Colour Press Ltd

Acknowledgements

The publishers wish to thank Cambridge Assessment International Education for permission to reproduce questions from past AS & A Level Mathematics papers. Cambridge Assessment International Education bears no responsibility for the example answers to questions taken from its past papers. These have been written by the authors. Exam-style questions and sample answers have been written by the authors.

The publishers wish to thank the following for permission to reproduce photographs. Every effort has been made to trace copyright holders and to obtain their permission for the use of copyright material. The publishers will gladly receive any information enabling them to rectify any error or omission at the first opportunity.

pvi Markus Gann/Shutterstock, p1 Serhii Krot/Shutterstock, p36 Mila Supinskaya Glashchenko/Shutterstock, p65 M2020/Shutterstock, p84 Markus Gann/Shutterstock, p121 ChameleonsEye/Shutterstock, p153 nd3000/Shutterstock, p175 Namthip Muanthongthae/Shutterstock, p224 Sergey Nivens/Shutterstock, p242 Likoper/Shutterstock.

Full worked solutions for all exercises, exam-style questions and past paper questions in this book available to teachers by emailing international.schools@harpercollins.co.uk and stating the book title.

CONTENTS

Introduction   **v**

1 – Algebra   **1**
- 1.1   The modulus of a linear function   2
- 1.2   Dividing polynomials   11
- 1.3   The factor theorem and the remainder theorem   14
- 1.4   Partial fractions   17
- 1.5   Binomial expansions   27

2 – Logarithms and exponential functions   **36**
- 2.1   Logarithms   37
- 2.2   Logarithms in other bases   41
- 2.3   The number e   44
- 2.4   Natural logarithms   49
- 2.5   Using logarithms to solve equations and inequalities   52
- 2.6   Logarithmic graphs   55

3 – Trigonometry   **65**
- 3.1   Addition and subtraction formulae   66
- 3.2   Double angle formulae   70
- 3.3   The expression $a \sin \theta + b \cos \theta$ 73
- 3.4   The secant, cosecant and tangent functions   76
- 3.5   More trigonometric identities   79

4 – Differentiation   **84**
- 4.1   Differentiating e^x 85
- 4.2   Differentiating $\ln x$ 88
- 4.3   Differentiating $\sin x$ and $\cos x$ 91
- 4.4   The product rule   94
- 4.5   The quotient rule   98
- 4.6   Differentiating $\tan^{-1} x$ 102
- 4.7   Differentiating parametric equations   105
- 4.8   Differentiating implicit equations   111

5 – Integration   **121**
-   The trapezium rule   122
- 5.2   Recognising integrals   127
- 5.3   Integration using trigonometrical relationships   132

- 5.4   Integrating $\dfrac{1}{x^2 + a^2}$ 134
- 5.5   Integrating expressions of the form $\dfrac{f'(x)}{f(x)}$ 136
- 5.6   Integration using partial fractions   139
- 5.7   Integration using a substitution   142
- 5.8   Integration by parts   145

6 – Numerical solution of equations   **153**
- 6.1   Finding roots   154
- 6.2   How change of sign methods can fail   158
- 6.3   Iterative methods   162
- 6.4   How iterative methods can vary   169

7 – Vectors   **175**
- 7.1   Definition of a vector   177
- 7.2   Vector geometry   182
- 7.3   Magnitude of a vector   189
- 7.4   Position vectors   194
- 7.5   Vector equation of a straight line   198
- 7.6   Parallel, intersecting and skew lines   201
- 7.7   Scalar product   205

8 – Differential equations   **224**
- 8.1   Constructing differential equations   226
- 8.2   Solving differential equations by integration   229
- 8.3   Separation of variables   234

9 – Complex numbers   **242**
- 9.1   Definition of a complex number   244
- 9.2   Addition, subtraction, multiplication and division of complex numbers   250
- 9.3   Complex roots of polynomial equations   256
- 9.4   Polar form   259
- 9.5   Geometric effects   264
- 9.6   Loci   270

Summary review   **284**

Glossary   **293**

Index   **296**

Blue highlighted sections are content that is only needed for Pure Mathematics 3.

Green highlighted sections are content that is only needed for Pure Mathematics 2.

Full worked solutions for all exercises, exam-style questions and past paper questions in this book available to teachers by emailing international.schools@harpercollins.co.uk and stating the book title.

INTRODUCTION

This book is part of a series of nine books designed to cover the content of the Cambridge International AS and A Level Mathematics and Further Mathematics. The chapters within each book have been written to mirror the syllabus, with a focus on exploring how the mathematics is relevant to a range of different careers or to further study. This theme of *Mathematics in life and work* runs throughout the series, with regular opportunities to deepen your knowledge through group discussion and exploring real-world contexts.

Within each chapter, examples are used to introduce important concepts and practice questions are provided to help you to achieve mastery. Developing skills in mathematical modelling, problem solving and communication can significantly strengthen overall mathematical ability. The practice questions in every chapter have been written with this in mind and selected questions include symbols to indicate which of these underlying skills are being developed. Exam-style questions are included at the end of each chapter and a bank of practice questions, including real Cambridge past exam questions, is included at the end of the book.

A range of other features throughout the series will help to optimise your learning. These include:

> **key information boxes** – highlighting important learning points or key formulae

> **commentary boxes** – tackling potential misconceptions and strengthening understanding through probing questions

> **stop and think** – encouraging independent thinking and developing reflective practice.

Key mathematical terminology is listed at the beginning of each chapter and a glossary is provided at the end of each book. Similarly, a summary of key points and key formulae is provided at the end of each chapter. Where appropriate, alternative solutions are included within the worked solutions to encourage you to consider different approaches to solving problems.

Pure Mathematics 2 and 3 will provide you with opportunities to become more fluent and confident with algebraic manipulation. In particular, you will learn new methods for differentiating and integrating composite and increasingly complex functions. You will be introduced to a very important irrational number, $e = 2.718\ldots$, which has many real-life applications, including calculating compound interest and exponential decay. You will encounter problems that can be solved using multiple methods and approaches as well as some problems that cannot be solved analytically and must be solved using a numerical approach.

In Pure Mathematics 3, you will build on your knowledge of vectors from your Upper Secondary course by expanding into 3D and using vector notation to represent straight lines. You will also be introduced to new areas of mathematics, including differential equations and complex numbers. The additional material for Pure Mathematics 3 is indicated by a coloured bar at the side of the page.

FEATURES TO HELP YOU LEARN

Mathematics in life and work

Each chapter starts with real-life applications of the mathematics you are learning in the chapter to a range of careers. This theme is picked up in group discussion activities throughout the chapter.

Learning objectives

A summary of the concepts, ideas and techniques that you will meet in the chapter.

Language of mathematics

Discover the key mathematical terminology you will meet in this chapter. As you work through the chapter, you will encounter key words, written in bold. The words are defined in the glossary at the back of the book.

Prerequisite knowledge

See what mathematics you should know before you start the chapter, with some practice questions to check your understanding.

---

4 DIFFERENTIATION

Mathematics in life and work

When quantities change over time, differentiation is the ideal mathematical tool to analyse and study those changes. Differentiation has a wide variety of applications across the physical and natural sciences – for example:

> If you were a molecular biologist, you would use differentiation to analyse changes taking place in a body when a drug is used to combat disease.

> If you were an industrial chemist, you would use differentiation to look at the rates at which chemical reactions take place in different circumstances.

> If you were an environmental scientist, you would use differentiation when studying changes in populations of animals or plants under different conditions.

> If you were an astronomer, you would use differentiation to study the motion of planets, stars and other astronomical bodies.

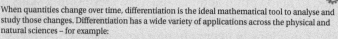

> If you were an engineer, you would use differentiation to analyse oscillations in any vibrating system, such as a car suspension or a bridge in the wind.

LEARNING OBJECTIVES

You will learn how to:

> differentiate e^x, $\ln x$, $\sin x$, $\cos x$, $\tan x$ and $\tan^{-1} x$

> differentiate products and quotients

> differentiate functions defined parametrically or implicitly.

LANGUAGE OF MATHEMATICS

Key words and phrases you will meet in this chapter:

> implicit equation, parametric equation, product rule, quotient, quotient rule

PREREQUISITE KNOWLEDGE

You should already know how to:

> differentiate x^n for any value of n

> find the equation of a tangent or a normal to a curve

Explanations and examples

Each section begins with an explanation and one or more worked examples, with commentary where appropriate to help you follow. Some show alternative solutions in the example of accompanying commentary to get you thinking about different approaches to a problem.

Example 5

Solve $|3x - 2| = |2x + 7|$.

Solution

Using the relation $|a| = |b| \Leftrightarrow a^2 = b^2$, you can rewrite this equation as

$(3x - 2)^2 = (2x + 7)^2$.

Expand the brackets.

$9x^2 - 12x + 4 = 4x^2 + 28x + 49$

Colour-coded questions

Questions are colour-coded (green, blue and red) to show you how difficult they are. Exercises start with more accessible (green) questions and then progress through intermediate (blue) questions to more challenging (red) questions.

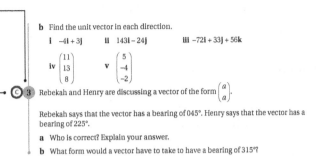

b Find the unit vector in each direction.

i $-4i + 3j$ **ii** $143i - 24j$ **iii** $-72i + 33j + 56k$

iv $\begin{pmatrix} 11 \\ 13 \\ 8 \end{pmatrix}$ **v** $\begin{pmatrix} 5 \\ -4 \\ -2 \end{pmatrix}$

3 Rebekah and Henry are discussing a vector of the form $\begin{pmatrix} a \\ a \end{pmatrix}$.

Rebekah says that the vector has a bearing of 045°. Henry says that the vector has a bearing of 225°.

a Who is correct? Explain your answer.

b What form would a vector have to take to have a bearing of 315°?

Question-type indicators

The key concepts of problem solving, communication and mathematical modelling underpin your A level Mathematics course. You will meet them in your learning throughout this book and they underpin the exercises and exam-style questions. All mathematics questions will include one or more of the key concepts in different combinations. We have labelled selected questions that are especially suited to developing one or more of these key skills with these icons:

(PS) Problem solving – mathematics is fundamentally about problem solving and representing systems and models in different ways. These include: algebra, geometrical techniques, calculus, mechanical models and statistical methods. This icon indicates questions designed to develop your problem-solving skills. You will need to think carefully about what knowledge, skills and techniques you need to apply to the problem to solve it efficiently.

These questions may require you to:

- use a multi-step strategy
- choose the most efficient method, or
- bring in mathematics from elsewhere in the curriculum
- look for anomalies in solutions
- generalise solutions to problems.

(C) Communication – communication of steps in mathematical proof and problem solving needs to be clear and structured, using algebra and mathematical notation, so that others can follow your line of reasoning. This icon indicates questions designed to develop your mathematical communication skills. You will need to structure your solution clearly, to show your reasoning and you may be asked to justify your conclusions.

These questions may require you to:

- use mathematics to demonstrate a line of argument
- make use of mathematical notation in your solution
- follow mathematical conventions to present your solution clearly
- justify why you have reached a conclusion.

(MM) Mathematical modelling – a variety of mathematical content areas and techniques may be needed to turn a real-world situation into something that can be interpreted through mathematics. This icon indicates questions designed to develop your mathematical modelling skills. You will need to think carefully about what assumptions you need to make to model the problem, and how you can interpret the results to give predictions and information about the real world.

These questions may require you to:

- construct a mathematical model of a real-life situation, using a variety of techniques and mathematical concepts

- » use your model to make predictions about the behaviour of mathematical systems
- » make assumptions to simplify and solve a complex problem.

Key information

These boxes highlight information that you
need to pay attention to and learn, such as
key formulae and learning points

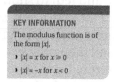

KEY INFORMATION

The modulus function is of
the form $|x|$.

- » $|x| = x$ for $x \geq 0$
- » $|x| = -x$ for $x < 0$

Stop and think

These boxes present you

Stop and think   What is the Cartesian equation of this circle?
Why might a parametric equation be more convenient?

with probing questions and problems to help you to reflect on what you have been learning.
They challenge you to think more widely and deeply about the mathematical concepts, tackle
misconceptions and, in some cases, generalise beyond the syllabus. They can be a starting point
for class discussions or independent research. You will need to think carefully about the question
and come up with your own solution.

Mathematics in life and work – Group discussions give you the chance to apply the
skills you have learned to a model of a real-life maths problem, from a career that uses
maths. Your focus is on applying and practising the concepts, and coming up with your own
solutions, as you would in the workplace. These tasks can be used for class discussions, group
work or as an independent challenge.

Summary of key points

At the end of each chapter, there is a summary of key formulae and learning points.

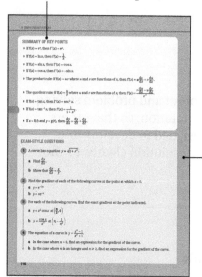

Exam-style questions

Practise what you have learnt throughout the chapter
with questions written in examination style by our
authors, progressing in order of difficulty.

The last **Mathematics in life and work** question draws
together the skills that you have gained in this chapter
and applies them to a simplified real-life scenario.

At the end of the book, test your mastery of what you have learned in the **Summary Review**
section. Practise the basic skills with some Cambridge International A Level Paper 1 questions,
and then go on to try carefully selected questions from Cambridge International A Level past exam
papers and exam-style questions on new topics. Extension questions, written by our authors, give
you the opportunity to challenge yourself and prepare you for more advanced study.

1 ALGEBRA

Mathematics in life and work

This chapter focuses on manipulating algebraic expressions to make them simpler. This could involve finding the factors of an expression so that it can be fully factorised, or splitting a function containing a fraction into partial fractions so that the function can be used. The mathematics used in this chapter is widely applicable in a range of careers – for example:

> If you were a designer of road traffic systems, you would need to make the distinction between velocity and speed. Algebra can be used to find scalar quantities from vector quantities.

> If you were a financier, you might analyse financial transactions to determine the total value of money moved. Algebra could be used to find only positive numerical values.

> If you were an engineer working on the Large Hadron Collider, you would need to analyse the movement of particles. Algebra could be used to model changes in mass and energy over time.

In this chapter, you will consider the application of algebra to the design of road traffic systems.

LEARNING OBJECTIVES

You will learn how to:

> understand the meaning of $|x|$, sketch the graph of $y = |ax + b|$ and use relations such as $|a| = |b| \Leftrightarrow a^2 = b^2$ and $|x - a| < b \Leftrightarrow a - b < x < a + b$ in the course of solving equations

> divide a polynomial by a linear or quadratic polynomial, and identify the quotient and remainder

> use the factor theorem and the remainder theorem

> recall an appropriate form for expressing rational functions in partial fractions, and carry out the decomposition

> use the expansion of $(1 + x)^n$, where n is a rational number and $|x| < 1$.

LANGUAGE OF MATHEMATICS

Key words and phrases you will meet in this chapter:

> absolute value, binomial expansion, divisor, factor theorem, modulus/moduli, partial fractions, polynomial, quotient, remainder, remainder theorem

PREREQUISITE KNOWLEDGE

You should already know how to:

- work with coordinates in all four quadrants
- recognise, sketch and interpret graphs of linear functions
- sketch transformations of a given function
- use and interpret algebraic manipulation
- use the concepts and vocabulary of expressions, equations, formulae, identities, inequalities, terms and factors
- simplify and manipulate algebraic expressions
- work out the coefficients in a binomial expansion.

You should be able to complete the following questions correctly:

1   Sketch the graphs of each of the following functions.

   a $y = 2x + 5$ **b** $y = 3x - 2$ **c** $y = 3 - x$

2   Expand and simplify the following expressions.

   a $(x + 1)(x + 2)$ **b** $(x - 3)(x + 4)$

   c $(x - 2)(x - 3)$ **d** $(x - 3)^2$

   e $(2x + 1)(x + 2)$ **f** $(x + 1)(x + 2)(x + 3)$

3   Find the coefficient of x^3 in a binomial expansion of $(2 + x)^6$.

1.1 The modulus of a linear function

The **modulus** of a number is its positive numerical value. The symbol for modulus is two vertical lines, one on either side of the function. For example, $|7| = 7$ and $|-7| = 7$. So you can see that any negative values are changed to be positive by the modulus.

Generally, the modulus function is of the form $|x|$.

$|x| = x$ for $x \geqslant 0$

$|x| = -x$ for $x < 0$

So all the values of $|x|$ for $x < 0$ are positive.

For example, the graphs of $y = x$ and $y = |x|$ are:

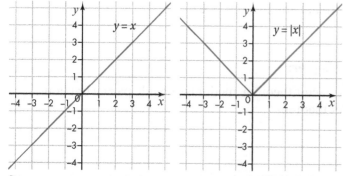

> On calculators, the modulus function is sometimes called **absolute value** so the button on your calculator may be labelled 'Mod' or 'Abs'.

> **KEY INFORMATION**
>
> The modulus function is of the form $|x|$.
>
> - $|x| = x$ for $x \geqslant 0$
> - $|x| = -x$ for $x < 0$

> When graphing modulus functions of the form $y = |f(x)|$, an alternative approach to the algebraic method is to reflect in the x-axis any part of the graph that appears below the x-axis.

Example 1

Sketch the graph of $y = |x - 5|$.

Solution

First, sketch the graph of $y = x - 5$.

Any negative values from the function (negative y-values) will be changed to positive by the modulus, a reflection in the x-axis.

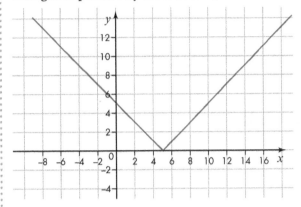

Example 2

Sketch the graph of $y = |x| - 5$.

Solution

First you need to sketch the graph of $y = x - 5$ for $x \geqslant 0$.

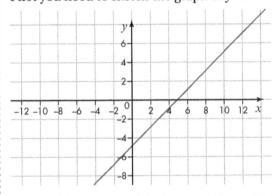

You then need to reflect the line in the y-axis.

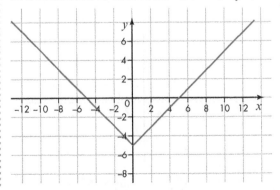

KEY INFORMATION

If you are not sure why the graph is reflected in the y-axis, create and compare the table of values for each of $y = x - 5$ and $y = |x| - 5$.

3

Alternatively you could draw the graph of $y = |x|$.

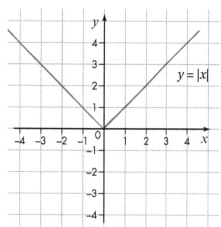

You then translate this graph by $\begin{pmatrix} 0 \\ -5 \end{pmatrix}$ to obtain the graph of

$y = |x| - 5$.

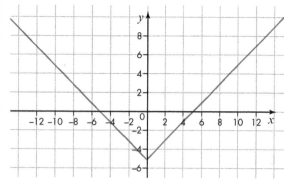

Exercise 1.1A

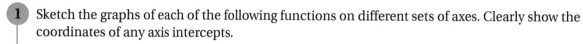

1 Sketch the graphs of each of the following functions on different sets of axes. Clearly show the coordinates of any axis intercepts.

a $y = |3x + 2|$ **b** $y = |x - 2|$

c $y = 3|x| + 2$ **d** $y = |x| - 2$

Ⓒ **2** Which of the following is the graph of $y = |2x - 5|$? Clearly explain and justify your choice.

A   **B**

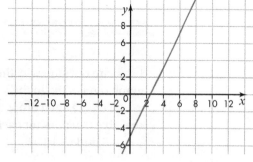 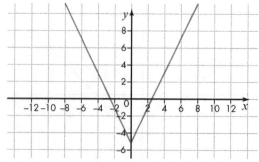

   Ⓒ **Communication**   ⓂⓂ **Mathematical modelling**   Ⓟ Ⓢ **Problem solving**

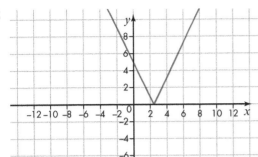

C

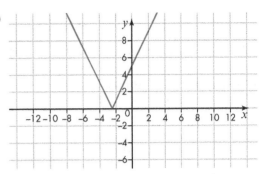

D

3 Sketch the graphs of each of the following functions on different sets of axes. Clearly show the coordinates of any axis intercepts.

    a $y = |6 - x|$ **b** $y = |-x|$

    c $y = 6 - |x|$ **d** $y = -|x|$

4 Here is the graph of $y = f(x)$.

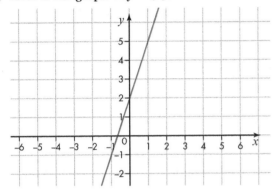

On separate sets of axes, sketch the graphs of each of the following functions. Clearly show the coordinates of any axis intercepts.

    a $y = |f(x)|$ **b** $y = f(|x|)$

PS **5** Given that $f(x) = 5 - 2x$, sketch the graph of $y = f(|x|)$.

PS **6** Given that $f(x) = 3 - 4x$, what transformation of this function would result in the function with the following graph?

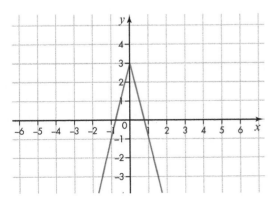

7 **a** Given that $f(x) = 3 - 4x$, sketch the graph of $f(|x + 2|)$.

b Describe the transformation from $f(x)$ to $f(|x + 2|)$.

8 Here is the graph of $y = f(x)$.

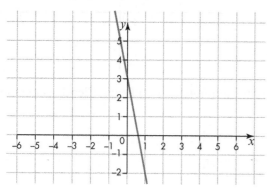

On separate pairs of axes, sketch the graphs of each of the following functions. Clearly show the coordinates of any axis intercepts.

a $y = f\left(\left|\dfrac{x}{2}\right|\right)$.

b $y = |f(x - 2)|$.

c $y = |f(x)| + 3$.

PS **9** Given that $f(x) = 2x - 7$, detail the transformation(s) of this function that would result in the function with the following graph.

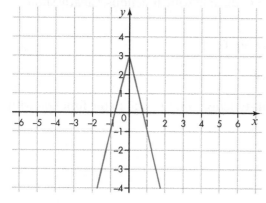

10 Given that $f(x) = 3x + \frac{1}{2}$ and that A with coordinates $\left(\frac{1}{2}, 2\right)$ is a point on this line, find the coordinates of A when the following transformations are applied.

a $y = f(|2x|)$ **b** $y = |f(x - 4)|$

c $y = |f(x)| + 5$

You need to be able to use the graphs of functions containing **moduli** to solve both equations and inequalities. For example, you need to be able to use the graph of $y = |2x - 1|$ to solve the equation $|2x - 1| = x$ or the inequality $|2x - 1| > x$.

> The plural of modulus is moduli.

Example 3

Solve the equation $|2x - 1| = x$.

Solution

On the same set of axes, sketch the graphs of $y = |2x - 1|$ and $y = x$.

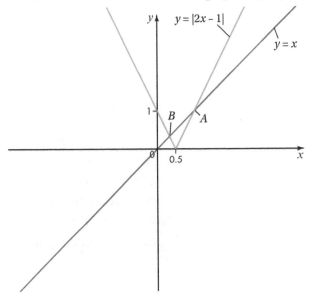

There are two points of intersection, A and B.

At A: $2x - 1 = x$

Solving gives

$x = 1$

At B: $-(2x - 1) = x$

Solving gives $-2x + 1 = x$

$x = \frac{1}{3}$

> An alternative algebraic method of solving this equation is to square both sides, which removes the modulus. You get:
>
> $(2x - 1)^2 = x^2$
>
> $4x^2 - 4x + 1 = x^2$
>
> $3x^2 - 4x + 1 = 0$
>
> $(3x - 1)(x - 1) = 0$
>
> $x = \frac{1}{3}$ or $x = 1$

Example 4

Solve the inequality $|2x - 1| > x$.

Solution

On the same set of axes, sketch the graphs of $y = |2x - 1|$ and $y = x$.

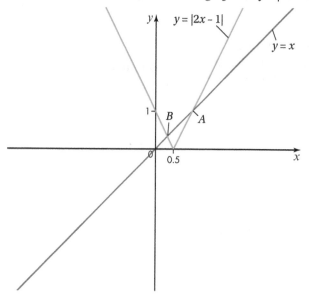

There are two points of intersection, A and B.

At A: $2x - 1 > x$

Solving gives $x > 1$

At B: $-(2x - 1) > x$

Solving gives $-2x + 1 > x$

$x < \dfrac{1}{3}$

So $x < \dfrac{1}{3}$ or $x > 1$.

> An alternative method is to solve this inequality by a visual inspection of the graph. You know from **Example 3** that at A, $x = 3$, and at B, $x = \dfrac{1}{3}$. $|2x - 1| > x$ when the green graph is above the red graph.
>
> This occurs when $x < \dfrac{1}{3}$ or when $x > 1$.

There are some useful relations that can be employed when solving equations and inequalities involving the modulus function.

Consider $a^2 = 36$ and $b^2 = 36$. What can you say about a and b?

If $a = -6$ and $b = 6$, then clearly a and b are not equal, and vice versa. However, if you use the modulus function, $|a| = 6$ which is equal to $|b| = 6$.

So you can say:

$|a| = |b| \Leftrightarrow a^2 = b^2$

> In this example, a could be -6 or $+6$ and b could also be -6 or $+6$.

This relation means if the squares of a and b are the same then the positive numerical values of a and b will also be the same.

The double-headed arrow $\Leftrightarrow$ means that if whatever is on the right-hand side of the arrow is true, then whatever is on the left-hand side of the arrow is true. It is also referred to as 'iff' or 'if and only if'.

Now look at a different example. Consider $x - 3 < 5$. This simplifies to $x < 8$. But what if $|x - 3| < 5$?

If you remove the modulus, then $x - 3$ could be between two values, -5 and 5. So you could rewrite the inequality as

$-5 < x - 3 < 5$.

$\therefore -2 < x < 8$

So you can say:

$|x - a| < b \Leftrightarrow a - b < x < a + b$

Here is a final example. Consider $x - 3 \geq 5$. This simplifies to $x \geq 8$. However, what if $|x - 3| \geq 5$?

If you remove the modulus, then $x - 3 \geq 5$ or $x - 3 \leq -5$.

Simplify by adding 3 to each side of the inequalities:

$x \leq -2$ or $x \geq 8$.

So you can say:

$|x - a| \geq b \Leftrightarrow x \leq a - b$ or $x \geq a + b$

KEY INFORMATION

$|a| = |b| \Leftrightarrow a^2 = b^2$

KEY INFORMATION

$|x - a| < b \Leftrightarrow a - b < x < a + b$

KEY INFORMATION

$|x - a| \geq b \Leftrightarrow x \leq a - b$ or $x \geq a + b$

You should be aware that the strictness of inequalities must be consistent on both sides of the equivalence sign ($\Leftrightarrow$).

Example 5

Solve $|3x - 2| = |2x + 7|$.

Solution

Using the relation $|a| = |b| \Leftrightarrow a^2 = b^2$, you can rewrite this equation as

$(3x - 2)^2 = (2x + 7)^2$.

Expand the brackets.

$9x^2 - 12x + 4 = 4x^2 + 28x + 49$

Rearrange.

$5x^2 - 40x - 45 = 0$

Divide by 5.

$x^2 - 8x - 9 = 0$

Factorise.

$(x - 9)(x + 1) = 0$

So $x = -1$ or $x = 9$.

An alternative method of solving this equation would have been to sketch the graphs of both modulus functions and find the x-coordinates of the points of intersection.

Example 6

Solve $|x + 1| > 2x + 5$.

Solution

Using the relation $|x - a| \geqslant b \Leftrightarrow x \leqslant a - b$ or $x \geqslant a + b$, you can rewrite this inequality as

$x + 1 > 2x + 5$ or $x + 1 < -(2x + 5)$.

Rearrange.

$x < -4$ or $x < -2$

> **Stop and think** It is important to use the word 'or' here. What difference would it have made if the word 'and' had been used?

So the solution is $x < -2$.

How could you check the solutions to **Example 5** and **Example 6**?

Exercise 1.1B

1　**a**　On the same set of axes, sketch the graphs of $y = |3x - 4|$ and $y = 2$.

　　b　Use your graphs to solve $|3x - 4| = 2$.

2　**a**　On the same set of axes, sketch the graphs of $y = |x - 5|$ and $y = 2x$.

　　b　Use your graphs to solve $|x - 5| > 2x$.

3　**a**　Using the relation $|a| = |b| \Leftrightarrow a^2 = b^2$, solve the equation $|2x - 1| = |1 - x|$.

　　b　On the same pair of axes, sketch the graphs of $y = |2x - 1|$ and $y = |1 - x|$ to verify your solutions to **part a**.

4　Using the relation $|x - a| < b \Leftrightarrow a - b < x < a + b$, solve the inequality $|2 - 5x| < 4x$.

(PS) 5　Solve $|1 - 3x| = |4x + 5|$.

(PS) 6　Solve $4 - x < |2x - 7|$.

7　Solve $|2x - 7| = |6 - x|$.

8　**a**　Describe the relation you would use to solve $|3 - 7x| > 2x$.

　　b　Solve $|3 - 7x| > 2x$.

9　Solve $|3x - 5| = |4 - 2x|$.

10　Given that $f(x) = 5 - 4x$ and $g(x) = 3x$, solve $|f(x)| > g(|x|)$.

Mathematics in life and work: Group discussion

As a designer of road traffic systems, you need to deal with scalar quantities (those with just magnitude, for example, distance) and vector quantities (those with both magnitude and direction, for example, displacement).

Two vehicles are travelling towards each other on different sides of the road. Vehicle A has a velocity of $13.4\,\mathrm{m\,s^{-1}}$ and vehicle B has a velocity of $-17.9\,\mathrm{m\,s^{-1}}$.

1 From the information you have been given, how do you know that the two vehicles are travelling in opposite directions to each other?

2 What is the difference between velocity and speed? How could you work out the speed of each vehicle in $\mathrm{km\,h^{-1}}$?

3 If the cars travel towards each other for $10\,\mathrm{s}$, what will be the displacement of each vehicle from its starting point? How will this differ from its distance travelled?

4 How would you represent each vehicle on a velocity–time graph? If the vehicles started accelerating towards each other, how would the velocity–time graph change?

1.2 Dividing polynomials

A method for dividing a **polynomial** by an expression is very similar to the method for numerical long division.

Example 7

Divide $x^3 + 6x^2 + 11x + 6$ by $(x + 2)$ and then fully factorise the expression.

Solution

Divide $x^3 + 6x^2 + 11x + 6$ by $(x + 2)$.

$$x + 2\,\overline{\smash{)}\,x^3 + 6x^2 + 11x + 6}$$

Divide x^3 by x.

$$\begin{array}{r} x^2 \\ x + 2\,\overline{\smash{)}\,x^3 + 6x^2 + 11x + 6} \end{array}$$

Multiply x^2 by $(x + 2)$.

$$\begin{array}{r} x^2 \\ x + 2\,\overline{\smash{)}\,x^3 + 6x^2 + 11x + 6} \\ x^3 + 2x^2 \end{array}$$

Work out the **remainder** by subtracting $x^3 + 2x^2$ from $x^3 + 6x^2$. Carry down the next part of the expression.

$$\begin{array}{r} x^2 \\ x + 2\,\overline{\smash{)}\,x^3 + 6x^2 + 11x + 6} \\ \underline{x^3 + 2x^2} \\ 0 + 4x^2 + 11x \end{array}$$

Divide $4x^2$ by x.

> It is critical that the sum is laid out correctly to avoid errors.

$$x + 2 \overline{\smash{\big)}\, \begin{array}{l} x^2 + 4x \\ x^3 + 6x^2 + 11x + 6 \end{array}}$$
$$\underline{x^3 + 2x^2}$$
$$0 + 4x^2 + 11x$$

Multiply $4x$ by $(x + 2)$.

$$x + 2 \overline{\smash{\big)}\, \begin{array}{l} x^2 + 4x \\ x^3 + 6x^2 + 11x + 6 \end{array}}$$
$$\underline{x^3 + 2x^2}$$
$$0 + 4x^2 + 11x$$
$$4x^2 + 8x$$

Work out the remainder by subtracting $4x^2 + 8x$ from $4x^2 + 11x$. Carry down the next part of the expression.

$$x + 2 \overline{\smash{\big)}\, \begin{array}{l} x^2 + 4x \\ x^3 + 6x^2 + 11x + 6 \end{array}}$$
$$\underline{x^3 + 2x^2}$$
$$0 + 4x^2 + 11x$$
$$\underline{4x^2 + 8x}$$
$$3x + 6$$

Divide $3x$ by x.

$$x + 2 \overline{\smash{\big)}\, \begin{array}{l} x^2 + 4x + 3 \\ x^3 + 6x^2 + 11x + 6 \end{array}}$$
$$\underline{x^3 + 2x^2}$$
$$0 + 4x^2 + 11x$$
$$\underline{4x^2 + 8x}$$
$$3x + 6$$

Multiply 3 by $(x + 2)$. Work out the remainder by subtracting $3x + 6$ from $3x + 6$. Remainder 0.

$$x + 2 \overline{\smash{\big)}\, \begin{array}{l} x^2 + 4x + 3 \\ x^3 + 6x^2 + 11x + 6 \end{array}}$$
$$\underline{x^3 + 2x^2}$$
$$0 + 4x^2 + 11x$$
$$\underline{4x^2 + 8x}$$
$$3x + 6$$
$$\underline{3x + 6}$$
$$0$$

So $x^3 + 6x^2 + 11x + 6 = (x + 2)(x^2 + 4x + 3)$.

Factorise the quadratic.

$(x + 2)(x^2 + 4x + 3) = (x + 2)(x + 1)(x + 3)$

Example 7 involved a cubic expression being divided by a linear one. You can also divide by higher degree polynomials, for example, quadratics. The process is no different.

KEY INFORMATION

The expression found after the algebraic division is known as the **quotient**.

Example 8

Divide $x^3 + 6x^2 + 11x + 6$ by $x^2 + 4x + 3$.

Solution

x^3 divided by x^2 is x; then multiply $x^2 + 4x + 3$ by x.

$$
\begin{array}{r}
x \\
x^2 + 4x + 3 \overline{\smash{\big)}\ x^3 + 6x^2 + 11x + 6} \\
x^3 + 4x^2 + 3x
\end{array}
$$

Work out the remainder.

$$
\begin{array}{r}
x \\
x^2 + 4x + 3 \overline{\smash{\big)}\ x^3 + 6x^2 + 11x + 6} \\
\underline{x^3 + 4x^2 + 3x } \\
2x^2 + 8x + 6
\end{array}
$$

Now start the process again by dividing $2x^2$ by x^2.

$$
\begin{array}{r}
x + 2 \\
x^2 + 4x + 3 \overline{\smash{\big)}\ x^3 + 6x^2 + 11x + 6} \\
\underline{x^3 + 4x^2 + 3x } \\
2x^2 + 8x + 6 \\
\underline{2x^2 + 8x + 6} \\
0
\end{array}
$$

Example 9

Divide $x^3 + 6x^2 + 11x + 6$ by $(x - 3)$ and find the remainder.

Solution

Divide $x^3 + 6x^2 + 11x + 6$ by $(x - 3)$.

$$
\begin{array}{r}
x^2 + 9x + 38 \\
x - 3 \overline{\smash{\big)}\ x^3 + 6x^2 + 11x + 6} \\
\underline{x^3 - 3x^2 } \\
0 + 9x^2 + 11x \\
\underline{9x^2 - 27x } \\
38x + 6 \\
\underline{38x - 114} \\
120
\end{array}
$$

The remainder is 120.
You could now express the original expression as follows:

$$x^3 + 6x^2 + 11x + 6 \equiv (x^2 + 9x + 38)(x - 3) + 120.$$

When finding the remainder, remember that you subtract the next row of working from the previous, for example, $x^3 - x^3 = 0$ and $6x^2 - -3x^2 = 9x^2$. Note that the double negative turns to a positive.

When dividing by a linear term $(x - a)$, a useful check is to substitute $x = a$ into the original expression. This should give the remainder. In this case, substituting $x = 3$ into the original expression gives 120. This is also a way of finding the remainder without having to complete the full process of long division. A fuller explanation of this is provided in **Example 11**.

Exercise 1.2A

1 Use algebraic division to show that, when $x^3 + 4x^2 + x - 6$ is divided by $(x - 1)$, the remainder is zero.

2 Find the remainder when $2x^3 + 15x^2 + 31x + 12$ is divided by $(x - 1)$.

3 Using algebraic division, determine if $(x + 1)$ is a factor of:

    a $3x^3 + 2x^2 - 7x + 2$ **b** $2x^4 + 3x^3 - 12x^2 - 7x + 6$.

4 Use algebraic division to find the remainder when:

    a $x^4 - 18x + 81$ is divided by $(x - 2)$ **b** $8x^4 + 2x^3 - 53x^2 + 37x - 6$ is divided by $(x + 1)$.

5 Use algebraic division to find the remainder when $3x^3 - 4x^2 + 1$ is divided by $(2x + 1)$.

(PS) 6 When $2x^3 + 5x^2 + px - q$ is divided by $(x - 2)$ or $(x + 4)$, the remainder is zero. Find the values of p and q.

(PS) 7 When $x^3 + 3x^2 - 2px + p$ is divided by $(x + 5)$, the remainder is -33. Find the value of p.

(PS) 8 When $x^3 + ax^2 - 13x + b$ is divided by $(x + 5)$, the remainder is 0. When $x^3 + bx^2 + ax + 10$ is divided by $(x - 1)$, the remainder is -1. Find the values of a and b.

(PS) 9 When the expressions $mx^3 + x^2 - 10x - 6$ and $2x^3 - 4x^2 + mx + 8$ are divided by $(x - 2)$, the remainders are the same. Find the value of m.

(PS) 10 Find the remainder, if any, when $5x^3 + 8x^2 - 41x + 28$ is divided by $x^2 + 3x - 4$. Give your answer in the form $(ax + b)$, where a and b are integers.

1.3 The factor theorem and the remainder theorem

The factor theorem

Sometimes it isn't easy to spot factors, particularly in higher-order polynomials such as $x^4 + x^3 + x^2 + x + 1$. However, if you substitute a value, a, for x into an expression and the value of the expression is zero, then $(x - a)$ is a factor of the expression. This result is called the **factor theorem**.

> **KEY INFORMATION**
>
> If $f(a) = 0$, then $(x - a)$ is a factor of $f(x)$.

Example 10

Given the expression $f(x) = x^3 + 6x^2 + 11x + 6$, show that $(x + 2)$ is a factor.

> **KEY INFORMATION**
>
> If $(x - a)$ is a factor of $f(x)$, then a is a root of $f(x) = 0$.

Solution

If $(x + 2)$ is a factor of $f(x)$, then $f(-2) = 0$.

Substitute $x = -2$ into $f(x)$.

$f(-2) = (-2)^3 + 6(-2)^2 + 11(-2) + 6$

$\quad = -8 + 24 - 22 + 6$

$\quad = 0$

$f(-2) = 0$ so $(x + 2)$ is a factor.

The remainder theorem

If you substitute a value, a, into an expression and the value of the expression is *not* zero then $(x - a)$ is *not* a factor of the expression. Additionally, the value of the expression after the substitution is the remainder when the expression is divided by $(x - a)$. This result is called the **remainder theorem**.

> **KEY INFORMATION**
>
> If $f(a) \neq 0$, then $(x - a)$ is not a factor of $f(x)$ and $f(a)$ is the remainder.

Example 11

Given the expression $f(x) = x^3 + 6x^2 + 11x + 6$, show that $(x - 3)$ is not a factor and find the remainder.

Solution

If $(x - 3)$ is not a factor of $f(x)$, then $f(3) \neq 0$.

Substitute $x = 3$ into $f(x)$.

$f(3) = (3)^3 + 6(3)^2 + 11(3) + 6$

$\quad = 27 + 54 + 33 + 6$

$\quad = 120$

$f(3) = 120$ so $(x - 3)$ is not a factor and the remainder is 120.

Also, the remainder theorem shows that any polynomial can be written as a product of its quotient multiplied by its **divisor** plus the remainder:

$F(x) \equiv Q(x) \times \text{divisor} + \text{remainder}$

where $F(x)$ is the original polynomial and $Q(x)$ is the quotient.

Alternatively, divide $x^3 + 6x^2 + 11x + 6$ by $(x - 3)$ using the remainder theorem.

When a cubic is divided by a linear expression, the quotient will be a quadratic and the remainder will be a constant.

$x^3 + 6x^2 + 11x + 6 \equiv (Ax^2 + Bx + C)(x - 3) + D$

By substituting different values for x into this identity you can find the values of A, B, C and D.

> **KEY INFORMATION**
>
> $F(x) \equiv Q(x) \times \text{divisor} + \text{remainder}$

Let $x = 3$

$27 + 54 + 33 + 6 = (9A + 3B + C) \times 0 + D$

$D = 120$

When $x = 3$, the whole expression $(Ax^2 + Bx + C)(x - 3)$ will be zero and so D can be found.

Let $x = 0$

$6 = (C)(-3) + 120$

$C = 38$

When $x = 0$, the whole expression $(Ax^2 + Bx)$ will be zero, so C can be found now the value of D is known.

Compare coefficients for x^3.

$1 = A$

Compare coefficients for x^2.

$6 = -3A + B$

$B = 9$

So $x^3 + 6x^2 + 11x + 6 \equiv (x^2 + 9x + 38)(x - 3) + 120$.

Or you could write this as $\dfrac{x^3 + 6x^2 + 11x + 6}{x - 3} \equiv x^2 + 9x + 38 + \dfrac{120}{x - 3}$.

> The value of the coefficient of x^3 on the left-hand side of the identity needs to be compared with the value of the coefficient of x^3 on the right-hand side of the identity.

> On the right-hand side of the identity, the coefficient of x^2 will come from $(Ax^2)(-3) + (Bx)(x)$.

Example 12

Divide $2x^3 - 3x^2 + 1$ by $(2x + 1)$ and then fully factorise the expression.

Solution

The cubic is missing an expression for x. We need to add $0x$ to the polynomial to ensure that the division is carried out correctly.

$$
\begin{array}{r}
x^2 - 2x + 1 \\
2x + 1 \overline{)\ 2x^3 - 3x^2 + 0x + 1} \\
\underline{2x^3 + x^2} \\
0 - 4x^2 + 0x \\
\underline{-4x^2 - 2x} \\
2x + 1 \\
\underline{2x + 1} \\
0
\end{array}
$$

So $2x^3 - 3x^2 + 1 = (2x + 1)(x^2 - 2x + 1)$.

Factorise the quadratic.

$(2x + 1)(x^2 - 2x + 1) = (2x + 1)(x - 1)^2$

Stop and think How do the factor theorem and the remainder theorem assist in the process to factorise higher-order polynomials fully: for example, $x^4 - 5x^3 + 5x^2 - 6$?

Exercise 1.3A

1 By substituting $x = 1$, show that $(x - 1)$ is a factor of $f(x) = x^3 + 4x^2 + x - 6$.

2 By substituting $x = 1$, show that $(x - 1)$ is not a factor of $f(x) = 2x^3 + 15x^2 + 31x + 12$ and state the remainder.

3 Using a combination of the factor theorem and algebraic division, fully factorise the following expressions.

a $3x^3 + 2x^2 - 7x + 2$

b $2x^4 + 3x^3 - 12x^2 - 7x + 6$

4 Use the remainder theorem to find the remainder when:

a $x^4 - 18x + 81$ is divided by $(x - 2)$

b $8x^4 + 2x^3 - 53x^2 + 37x - 6$ is divided by $(x + 1)$.

5 Show that $(2x + 1)$ is not a factor of $3x^3 - 4x^2 + 1$.

6 **a** Fully factorise $x^4 - 7x^3 + 13x^2 + 3x - 18$.

b Solve $x^4 - 7x^3 + 13x^2 + 3x - 18 = 0$.

7 **a** Using the factor theorem, show that $(x + 1)$ is not a factor of $2x^3 - 13x^2 - 10x + 21$.

b Using the factor theorem, fully factorise $2x^3 - 13x^2 - 10x + 21$.

c Solve $2x^3 - 13x^2 - 10x + 21 = 0$.

8 **a** Using the factor theorem, show that $3x^3 - 5x^2 - 47x - 15$ can be fully factorised.

b Solve $5x^3 - 15x^2 - 47x - 15 = 2x^3 - 10x^2$.

9 **a** Using the factor theorem fully factorise $2x^3 + 9x^2 - 20x - 12$.

b Solve $2x^3 + 9x^2 = 20x + 12$.

10 Using the factor theorem show that $12x^2 + 5x + 7$ cannot be factorised. You must justify your answer.

1.4 Partial fractions

Partial fractions without repeated factors

In your previous study, you simplified expressions such as $\frac{1}{x-1} + \frac{2}{x+1}$. What steps would you have needed to taken?

First, you would have found a common denominator.

$$= \frac{(x+1) + 2(x-1)}{(x-1)(x+1)}$$

Then you would have simplified the numerator and expanded the brackets in the denominator.

$$= \frac{x+1+2x-2}{x^2-1}$$

Then you would have collected like terms in the numerator.

$$= \frac{3x-1}{x^2-1}$$

When you split a fraction into **partial fractions** you are doing the reverse of the above. So, to split $\frac{3x-1}{x^2-1}$ into partial fractions, you

would have $\frac{3x-1}{x^2-1} \equiv \frac{1}{x-1} + \frac{2}{x+1}$

KEY INFORMATION

Partial fractions are two or more fractions into which a more complex fraction can be split.

The identity sign $\equiv$ has been used in the preceding text, rather than an equals sign, to show that $\dfrac{1}{x-1} + \dfrac{2}{x+1}$ is the same as $\dfrac{3x-1}{x^2-1}$ but is just written in a different way.

An expression with two or more linear terms in the denominator, for example, $\dfrac{f(x)}{(x+p)(x+q)(x+r)}$ can be split into partial fractions in the form $\dfrac{A}{x+p} + \dfrac{B}{x+q} + \dfrac{C}{x+r}$, where A, B and C are constants.

> **KEY INFORMATION**
>
> $$\dfrac{f(x)}{(x+p)(x+q)(x+r)}$$
>
> $$\equiv \dfrac{A}{x+p} + \dfrac{B}{x+q} + \dfrac{C}{x+r}$$

Example 13

Use substitution to split $\dfrac{2x-13}{(x+1)(x-2)}$ into partial fractions.

Solution

Using constants as the numerators, write the expression as partial fractions.

$$\dfrac{2x-13}{(x+1)(x-2)} \equiv \dfrac{A}{x+1} + \dfrac{B}{x-2}$$

You now need to find the values of A and B. Add the two fractions on the right-hand side of the identity by finding a common denominator.

$$\dfrac{2x-13}{(x+1)(x-2)} \equiv \dfrac{A(x-2) + B(x+1)}{(x+1)(x-2)}$$

As the denominators on each side of the identity are the same, the numerators are also the same.

$2x - 13 \equiv A(x-2) + B(x+1)$

The values of constants A and B can be determined by finding and solving simultaneous equations.

Substitute $x = 2$.

$x = 2$ has been chosen so that you are left with an equation with only one unknown, which can then be found.

$2 \times 2 - 13 = B(2+1)$

$-9 = 3B$

$B = -3$

Substitute $x = -1$.

$2 \times (-1) - 13 = A(-1-2)$

$-15 = -3A$

$A = 5$

State the original fraction as partial fractions.

$$\therefore \dfrac{2x-13}{(x+1)(x-2)} \equiv \dfrac{5}{x+1} - \dfrac{3}{x-2}$$

If you are asked to 'split a fraction into partial fractions' or similar, you must state what the partial fractions are. If you are asked to 'find the values of A and B' or similar, it isn't necessary to state the partial fractions.

Example 14

Split $\dfrac{36 + 5x}{16 - x^2}$ into partial fractions by equating coefficients.

Solution

Write the expression, using constants as the numerators, as partial fractions.

$$\frac{36 + 5x}{16 - x^2} \equiv \frac{A}{4 - x} + \frac{B}{4 + x}$$

> Notice the difference of two squares here to split the denominator.

You now need to find the values of A and B. Add the two fractions on the right-hand side of the identity by finding a common denominator.

$$\frac{36 + 5x}{16 - x^2} \equiv \frac{A(4 + x) + B(4 - x)}{(4 - x)(4 + x)}$$

As the denominators on each side of the identity are the same, the numerators are also the same.

$$36 + 5x \equiv A(4 + x) + B(4 - x)$$

Expand the brackets on the right-hand side of the identity.

$$36 + 5x \equiv 4A + Ax + 4B - Bx$$

Collect like terms.

$$36 + 5x \equiv 4(A + B) + x(A - B)$$

Equate the constant terms.

$$4A + 4B = 36 \qquad \textcircled{1}$$

> You now have the first of two simultaneous equations where the constant elements of $36 + 5x \equiv 4(A + B) + x(A - B)$ have been equated and consequently are the same.

Equate the coefficients of x.

$$A - B = 5 \qquad \textcircled{2}$$

> You now have the second of two simultaneous equations where the coefficient of x of $36 + 5x \equiv 4(A + B) + x(A - B)$ have been equated and consequently are the same.

Equations $\textcircled{1}$ and $\textcircled{2}$ now need to be solved simultaneously. Multiply $\textcircled{2}$ by 4.

$$4A - 4B = 20 \qquad \textcircled{3}$$

Add $\textcircled{1}$ and $\textcircled{3}$.

> You can choose either method – substitution or equating coefficients – to split a fraction into partial fractions.

$$8A = 56$$

$$A = 7$$

Substitute into $\textcircled{2}$.

$$7 - B = 5$$

$$B = 2$$

State what the original fraction is as partial fractions.

$$\therefore \frac{36 + 5x}{16 - x^2} \equiv \frac{7}{4 - x} + \frac{2}{4 + x}$$

Example 15

Split $\dfrac{3x^2 + 2x - 1}{(x-1)(x-2)(x-3)}$ into partial fractions using substitution.

Solution

Write the expression, using constants as the numerators, as partial fractions.

$$\frac{3x^2 + 2x - 1}{(x-1)(x-2)(x-2)} \equiv \frac{A}{x-1} + \frac{B}{x-2} + \frac{C}{x-3}$$

You now need to find the values of A, B and C. Add the three fractions on the right-hand side of the identity by finding a common denominator.

$$\frac{3x^2 + 2x - 1}{(x-1)(x-2)(x-3)} \equiv \frac{A(x-2)(x-3) + B(x-1)(x-3) + C(x-1)(x-2)}{(x-1)(x-2)(x-3)}$$

As the denominators on each side of the identity are the same, the numerators are also the same.

$$3x^2 + 2x - 1 \equiv A(x-2)(x-3) + B(x-1)(x-3) + C(x-1)(x-2)$$

Substitute $x = 1$.

$$3 + 2 - 1 = A(-1)(-2)$$
$$A = 2$$

Substitute $x = 2$.

$$12 + 4 - 1 = B(1)(-1)$$
$$B = -15$$

Substitute $x = 3$.

$$27 + 6 - 1 = C(2)(1)$$
$$C = 16$$

State the original fraction as partial fractions.

$$\therefore \frac{3x^2 + 2x - 1}{(x-1)(x-2)(x-3)} \equiv \frac{2}{x-1} - \frac{15}{x-2} + \frac{16}{x-3}$$

Exercise 1.4A

1. Express each of the following as partial fractions:
   i using substitution   **ii** equating coefficients.

   a $\dfrac{2x - 5}{(x+2)(x+3)}$ **b** $\dfrac{6x - 12}{(x-1)(x+5)}$ **c** $\dfrac{3 - 5x}{(x-3)(x-7)}$

2 Show that $\dfrac{11-2x}{(x-2)(x+5)}$ can be written as $\dfrac{A}{x-2}+\dfrac{B}{x+5}$.

State the values of A and B.

(PS) 3 Express each of the following as partial fractions.

a $\dfrac{2-3x}{(2x+1)(3-x)}$

b $\dfrac{2x+22}{x^2+2x}$

c $\dfrac{4x-30}{x^2-8x+15}$

(C) 4 Express $\dfrac{1}{x^2-9}$ and $\dfrac{1}{x^2-16}$ as partial fractions. What do you notice?

5 Express each of the following as partial fractions.

a $\dfrac{2x^2-4x+8}{(x-1)(x-2)(x-3)}$

b $\dfrac{2-3x-4x^2}{(x)(x-1)(1-2x)}$

c $\dfrac{6-6x-5x^2}{(x-1)(x-2)(x+4)}$

6 Show that $\dfrac{1}{x^2-a^2}$ can be written as $\dfrac{1}{2a(x-a)}-\dfrac{1}{2a(x+a)}$, where $a \in \mathbb{R}$.

(C) 7 Express $\dfrac{5+3x-x^2}{-x^3+3x^2+4x-12}$ as partial fractions. Provide a detailed commentary for each step in your working, ensuring that you clearly explain and justify any decisions you make.

8 Show that $\dfrac{5}{x^2-x+10}$ cannot be split into partial fractions. You must justify your answer.

9 Show that $\dfrac{x^2+x-6}{x^3+5x^2+2x-8}$ can be split into partial fractions.

10 Find the coefficient of $\dfrac{1}{2x+1}$ when $\dfrac{1}{2x^3-3x^2-32x-15}$ is split into partial fractions.

Partial fractions with repeated factors

An expression that has repeated linear factors in the denominator, for example $\dfrac{f(x)}{(x+p)(x+q)^2}$, can be split into partial fractions in the form $\dfrac{A}{x+p}+\dfrac{B}{x+q}+\dfrac{C}{(x+q)^2}$, where A, B and C are constants.

KEY INFORMATION

$$\dfrac{f(x)}{(x+p)(x+q)^2}$$

$$\equiv \dfrac{A}{x+p}+\dfrac{B}{x+q}+\dfrac{C}{(x+q)^2}$$

PURE MATHEMATICS 3

Example 16

Split $\dfrac{3x^2 + 2x + 2}{(x-2)(x-3)^2}$ into partial fractions.

Solution

Write the expression, using constants as the numerators, as partial fractions.

$$\frac{3x^2 + 2x + 2}{(x-2)(x-3)^2} \equiv \frac{A}{x-2} + \frac{B}{x-3} + \frac{C}{(x-3)^2}$$

You now need to find the values of A, B and C. Add the three fractions on the right-hand side of the identity by finding a common denominator.

$$\frac{3x^2 + 2x + 2}{(x-2)(x-3)^2} \equiv \frac{A(x-3)^2 + B(x-2)(x-3) + C(x-2)}{(x-2)(x-3)^2}$$

As the denominators on each side of the identity are the same, the numerators are also the same.

$3x^2 + 2x + 2 \equiv A(x-3)^2 + B(x-2)(x-3) + C(x-2)$

Expand the brackets on the right-hand side of the identity.

$3x^2 + 2x + 2 \equiv Ax^2 - 6Ax + 9A + Bx^2 - 5Bx + 6B + Cx - 2C$

Collect like terms.

$3x^2 + 2x + 2 \equiv x^2(A + B) + x(-6A - 5B + C) + 9A + 6B - 2C$

Equate the constant terms.

$9A + 6B - 2C = 2$ ①

Equate the coefficients of x.

$-6A - 5B + C = 2$ ②

Equate the coefficients of x^2.

$A + B = 3$ ③

Rearrange ③ to make B the subject.

$B = 3 - A$

Substitute B into ① and ②.

$9A + 6(3 - A) - 2C = 2$

$3A - 2C = -16$ ④

$-6A - 5(3 - A) + C = 2$

$-A + C = 17$ ⑤

Equations ④ and ⑤ now need to be solved simultaneously. Multiply ⑤ by 3.

$-3A + 3C = 51$ ⑥

> In this example, the method of equating coefficients has been used. An alternative method would have been to use substitution, as demonstrated in Example 15.

Add ④ and ⑥.

$$C = 35$$

Substitute into ⑤.

$$A = 18$$

Substitute into ③.

$$B = -15$$

State the original fraction as partial fractions.

$$\therefore \frac{3x^2 + 2x + 2}{(x-2)(x-3)^2} \equiv \frac{18}{x-2} - \frac{15}{x-3} + \frac{35}{(x-3)^2}$$

Exercise 1.4B

1 Show that $\dfrac{x^2 + 8x + 4}{x^2(x-2)}$ can be written as $\dfrac{A}{x} + \dfrac{B}{x^2} + \dfrac{C}{x-2}$. State the values of A, B and C.

(PS) 2 Given that $\dfrac{5x}{(x+3)^2} \equiv \dfrac{p}{x+3} - \dfrac{3p}{(x+3)^2}$. Find the value of p.

3 Express $\dfrac{7x - 3}{x^2 - 8x + 16}$ as partial fractions.

(PS) 4 Identify and correct the mistakes in the following to split $\dfrac{2x^2 - x - 6}{x^3 + 4x^2 + 4x}$ into partial fractions.

$$\frac{2x^2 - x - 6}{x^3 + 4x^2 + 4x} = \frac{2x^2 - x - 6}{x(x+2)^2}$$

$$\frac{2x^2 - x - 6}{x^3 + 4x^2 + 4x} \equiv \frac{A}{(x)} + \frac{B}{(x+2)} + \frac{C}{(x+2)^2}$$

$$\frac{2x^2 - x - 6}{x^3 + 4x^2 + 4x} \equiv \frac{A(x+2)^2 + B(x)(x+2) + C(x)}{x(x+2)^2}$$

$$2x^2 - x - 6 \equiv A(x+2)^2 + B(x)(x+2) + C(x)$$

Substitute $x = 0$.

$$-6 = 4A$$

$$A = -\frac{3}{2}$$

Substitute $x = -2$.

$$8 + 2 - 6 = -2C$$

$$C = -2$$

PURE MATHEMATICS 3

Substitute $x = 1$.

$$2 - 1 - 6 = \left(-\frac{3}{2}\right)(9) + B(1)(3) + (-2)(1)$$

$$B = \frac{7}{2}$$

$$\therefore \frac{2x^2 - x - 6}{x^3 + 4x^2 + 4x} \equiv -\frac{3}{x} + \frac{7}{x+2} + \frac{2}{(x+2)^2}$$

(PS) (C) **5** Can $\dfrac{1}{(x+1)(x-2)^2}$ be split into partial fractions? You must clearly justify your answer.

(PS) **6** Write down each step to express $\dfrac{2x^2 + 6x - 5}{(x-2)^3}$ as a partial fraction.

(C) **7** By choosing suitable examples to support your arguments, discuss the advantages and disadvantages of the two different methods for splitting a rational function with linear factors in its denominator into partial fractions.

8 Split $\dfrac{1}{x^3 - 6x^2 + 9x}$ into partial fractions.

9 Show that $\dfrac{2x-1}{(2x+1)^3}$ can be split into partial fractions.

10 Find the coefficient of $\dfrac{1}{(x+5)^2}$ when $\dfrac{x^2 + 7x - 1}{(x+5)^3}$ is split into partial fractions.

Partial fractions with quadratic factors

For each and every quadratic factor in the denominator (in the form $cx^2 + d$, where $c \geqslant 1$), there will be a related partial fraction in the form $\dfrac{Ax + B}{cx^2 + d}$, where A and B are constants.

> **KEY INFORMATION**
>
> For each and every quadratic factor in the denominator, there will be a related partial fraction in the form $\dfrac{Ax + B}{cx^2 + d}$.

Example 17

Split $\dfrac{x^2 - 3x + 4}{(x+1)(x^2 - 2)}$ into partial fractions.

Solution

Write the expression, using constants as the numerators, as partial fractions.

$$\frac{x^2 - 3x + 4}{(x+1)(x^2 - 2)} \equiv \frac{A}{x+1} + \frac{Bx + C}{x^2 - 2}$$

You now need to find the values of A, B and C. Add the two fractions on the right-hand side of the identity by finding a common denominator.

$$\frac{x^2 - 3x + 4}{(x + 1)(x^2 - 2)} \equiv \frac{A(x^2 - 2) + (Bx + C)(x + 1)}{(x + 1)(x^2 - 2)}$$

As the denominators on each side of the identity are the same, the numerators are also the same.

$$x^2 - 3x + 4 \equiv A(x^2 - 2) + (Bx + C)(x + 1)$$

Let $x = -1$.

$1 + 3 + 4 = -A$

So $A = -8$.

Let $x = 0$.

$4 = (-8)(-2) + C$

So $C = -12$.

Let $x = 1$.

$1 - 3 + 4 = (-8)(-1) + (B - 12)(2)$

$2 = 8 + 2B - 24$

So $B = 9$.

State the original fraction as partial fractions.

$$\therefore \frac{x^2 - 3x + 4}{(x + 1)(x^2 - 2)} \equiv \frac{3(3x - 4)}{x^2 - 2} - \frac{8}{x + 1}$$

Stop and think   The denominator in this question, $(x + 1)(x^2 - 2)$, could have been written as three linear terms $(x + 1)(x + \sqrt{2})(x - \sqrt{2})$. How might this solution have differed from the example shown?

Exercise 1.4C

1   Show that $\dfrac{x^2 - 5x + 6}{(x + 2)(x^2 + 1)}$ can be written as $\dfrac{Ax + B}{x^2 + 1} + \dfrac{C}{x + 2}$. State the values of A, B and C.

(PS) 2   Identify and correct the mistakes in the following, to split $\dfrac{x^2 - 7x + 6}{(x + 3)(x^2 + 1)}$ into partial fractions.

$$\frac{x^2 - 7x + 6}{(x + 3)(x^2 + 1)} \equiv \frac{A}{x + 3} + \frac{Bx + C}{x^2 + 1}$$

$$\frac{x^2 - 7x + 6}{(x + 3)(x^2 + 1)} \equiv \frac{A(x^2 + 1) + (Bx + C)(x + 3)}{(x + 3)(x^2 + 1)}$$

$x^2 - 7x + 6 \equiv A(x^2 + 1) + (Bx + C)(x + 3)$

Let $x = -3$.

$9 + 21 + 6 = 10A$

So $A = \dfrac{18}{5}$.

Let $x = 0$.

$6 = \left(\dfrac{18}{5}\right) + 3C$

So $C = \dfrac{4}{5}$.

Let $x = 1$.

$1 - 7 + 6 = \left(\dfrac{18}{5}\right)(2) + \left(B + \dfrac{4}{5}\right)(4)$

$0 = \dfrac{36}{5} + 4B + \dfrac{16}{5}$

So $B = -\dfrac{13}{5}$.

State the original fraction as partial fractions.

$\therefore \dfrac{x^2 - 7x + 6}{(x + 3)(x^2 + 1)} \equiv \dfrac{18}{x + 3} + \dfrac{4 - 13x}{x^2 + 1}$

3 Express $\dfrac{(x - 3)^2}{x(x^2 - 6)}$ as partial fractions.

(PS) **4** Given that $\dfrac{x^2 + 8x + 7}{(x - 1)^2(x^2 + 2)} \equiv \dfrac{px - 37}{9(x^2 + 2)} - \dfrac{p}{9(x - 1)} + \dfrac{8p}{3(x - 1)^2}$. Find the value of p.

(PS) **5** Can $\dfrac{x^2 - 5}{x(x^2 - 3)}$ be split into partial fractions? You must clearly justify your answer.

(C)

(PS) **6** Write down each step to express $\dfrac{x^2 + 3x - 5}{(x - 1)(x^2 + 3)}$ as a partial fraction.

7 Show that $\dfrac{3x - 1}{(x + 5)(x^2 - 1)}$ can be split into partial fractions.

8 Split $\dfrac{3x + 2}{(2x - 5)(x^2 + 5)}$ into partial fractions.

9 Show that $\dfrac{2x + 1}{(2x + 3)(x^2 + 1)}$ can be split into partial fractions.

10 Find the coefficient of $\dfrac{1}{x^2 + 2}$ when $\dfrac{1}{(x^2 + 2)(2x - 1)}$ is split into partial fractions.

Mathematics in life and work: Group discussion

As a road traffic system designer, you have been given data about various journeys along a stretch of motorway. You need to analyse the journeys of various vehicles. For one particular vehicle, you are given the velocity–time graph for its journey along a very busy section of the motorway. Consequently, the velocity and acceleration of the vehicle are constantly changing. A graph of velocity against time for the journey is plotted and given by the function f(x).

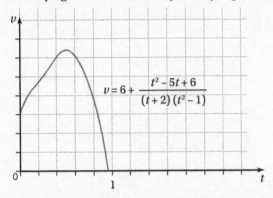

1   By analysing the graph, describe the journey of the vehicle. From the graph, how could you work out how far the vehicle has travelled?

2   The journey of the vehicle closely follows the curve given by $v = 6 + \dfrac{t^2 - 5t + 6}{(t+2)(t^2 - 1)}$.

To work out the distance travelled, the function will need to be integrated. Do you see this as being problematic? If so, why?

3   How can you integrate equations like the one in **Question 2**?

1.5 Binomial expansions

You encountered **binomial expansions** in **Pure Mathematics 1 Chapter 5 Series**.

$$(a + b)^n = a^n + na^{n-1}b + \frac{n(n-1)}{2}a^{n-2}(b)^2 + \ldots + (b)^n,$$

where n is a positive integer.

Put $a = 1$, $b = x$ and it becomes

$$(1 + x)^n = 1 + nx + \frac{n(n-1)}{2}x^2 + \frac{n(n-1)(n-2)}{3!}x^3 \ldots \qquad \text{①}$$

This is correct if n is a positive integer.

For example, if $n = 7$, it becomes $(1 + x)^7 = 1 + 7x + 21x^2 + \ldots + x^7$.

The series terminates if n is a positive integer.

In fact, you can put any rational non-integer or any negative integer into formula ① and it gives an infinite series.

Here is a general result:

$$(1 + x)^n = 1 + nx + \frac{n(n-1)}{2}x^2 + \frac{n(n-1)(n-2)}{3!}x^3 \ldots \text{ if } n \text{ is a}$$

rational number.

The series will converge if $|x| < 1$.

For example, if $n = -1$, you get the series

$$(1 + x)^{-1} = 1 + (-1)x + \frac{(-1)(-2)}{2}x^2 + \frac{(-1)(-2)(-3)}{3!}x^3 \ldots \text{ for } |x| < 1.$$

Tidy up the minus signs.

$$\frac{1}{1+x} = 1 - x + \frac{1 \times 2}{2}x^2 - \frac{1 \times 2 \times 3}{3!}x^3 + \ldots$$

All the coefficients cancel to 1.

$$\frac{1}{1+x} = 1 - x + x^2 - x^3 + \ldots$$

You should recognise the series on the right as a geometric progression with $a = 1$ and $r = -x$, which converges if $|x| < 1$.

As another example, try $n = \frac{1}{2}$. This gives the series

$$(1 + x)^{\frac{1}{2}} = 1 + \frac{1}{2}x + \frac{\frac{1}{2}\left(\frac{1}{2}-1\right)}{2}x^2 + \frac{\frac{1}{2}\left(\frac{1}{2}-1\right)\left(\frac{1}{2}-2\right)}{3!}x^3 \ldots$$

> The coefficient of x^3 is
> $$\frac{\frac{1}{2}\times\left(-\frac{1}{2}\right)\times\left(-\frac{3}{2}\right)}{6} = \frac{3}{8 \times 6} = \frac{1}{16}$$

$$\sqrt{1 + x} = 1 + \frac{1}{2}x - \frac{1}{8}x^2 + \frac{1}{16}x^3 \ldots |x| < 1.$$

Show that the next term in the binomial series for $\sqrt{1 + x}$ is

$$-\frac{5}{128}x^4.$$

Calculators do not 'know' square roots or any other function. They work them out by using series approximations of the type you have found here.

That is one reason why series approximations are so useful.

Here are a few binomial approximations.

You have already seen that $\frac{1}{1+x} = 1 - x + x^2 - x^3 + \ldots$.

If you replace x with $-x$, you get $\frac{1}{1-x} = 1 + x + x^2 + x^3 + \ldots$.

Putting $n = -\frac{1}{2}$ gives $\frac{1}{\sqrt{1+x}} = 1 - \frac{1}{2}x + \frac{3}{8}x^2 - \frac{5}{16}x^3 - \ldots$.

> **KEY INFORMATION**
>
> $$(1 + x)^n = 1 + nx + \frac{n(n-1)}{2}x^2$$
> $$+ \frac{n(n-1)(n-2)}{3!}x^3 \ldots \text{ if } n \text{ is a}$$
>
> rational number and $|x| < 1$. This is the general binomial expansion.

The original form of the binomial expansion (where n is a positive integer) is valid for any value of x and it produces a finite sequence of terms ($n + 1$ of them).

This new form gives an infinite sequence that will converge if $|x| < 1$.

Always give the coefficients as exact fractions rather than approximate decimals.

Example 20

a Work out a binomial series for $\dfrac{1}{\sqrt[3]{1-x}}$.

Give the terms up to x^3 and state the range of values of x for which it is valid.

b Use your series to find an approximate value for $\dfrac{1}{\sqrt[3]{7.2}}$.

c Use your answer to **a** to find a binomial series for $\dfrac{1}{\sqrt[3]{8-x}}$

and state the range of values for x for which it is valid.

Solution

a $\dfrac{1}{\sqrt[3]{1-x}} = (1-x)^{-\frac{1}{3}}$

$= 1 + \left(-\dfrac{1}{3}\right) \times (-x) + \dfrac{\left(-\frac{1}{3}\right) \times \left(-\frac{4}{3}\right)}{2}(-x)^2 + \dfrac{\left(-\frac{1}{3}\right) \times \left(-\frac{4}{3}\right) \times \left(-\frac{7}{3}\right)}{6}(-x)^3 + \ldots$

$= 1 + \dfrac{1}{3}x + \dfrac{2}{9}x^2 + \dfrac{14}{81}x^3 + \ldots$

The series is valid if $|-x| < 1$, which is the same as $|x| < 1$.

b $\dfrac{1}{\sqrt[3]{7.2}} = \dfrac{1}{\sqrt[3]{8-0.8}}$

To use a binomial expansion, you must have 1 instead of 8 so take out 8 as a factor.

$\dfrac{1}{\sqrt[3]{8-0.8}} = \dfrac{1}{\sqrt[3]{8(1-0.1)}}$

$= \dfrac{1}{2 \times \sqrt[3]{1-0.1}}$

$= \dfrac{1}{2} \times \dfrac{1}{\sqrt[3]{1-0.1}}$

Use the expansion from **part a** with $x = 0.1$.

$\dfrac{1}{\sqrt[3]{7.2}} \approx \dfrac{1}{2}\left(1 + \dfrac{1}{3} \times 0.1 + \dfrac{2}{9} \times 0.01 + \dfrac{14}{81} \times 0.001\right) = 0.517\,864$

to 6 d.p.

Note that this is close to the exact answer of 0.517 872 to 6 d.p.

c The expression under the cube root sign must start with 1 to form a binomial expansion.

Write $\dfrac{1}{\sqrt[3]{8-x}}$ as $\dfrac{1}{\sqrt[3]{8(1-\frac{1}{8}x)}} = \dfrac{1}{2}\left(1 - \dfrac{1}{8}x\right)^{-\frac{1}{3}}$.

Replace x in the series in **part a** by $\frac{1}{8}x$.

$$\frac{1}{\sqrt[3]{8-x}} = \frac{1}{2}\left\{1 + \frac{1}{3}\times\frac{1}{8}x + \frac{2}{9}\times\frac{1}{64}x^2 + \frac{14}{81}\times\frac{1}{512}x^3 + \ldots\right\}$$

$$= \frac{1}{2} + \frac{1}{48}x + \frac{1}{576}x^2 + \frac{7}{41472}x^3 + \ldots$$

The series is valid when $\left|\frac{1}{8}x\right| < 1$, which means $|x| < 8$.

Exercise 1.5A

1 **a** Work out the first three terms in the binomial expansion of $\dfrac{1}{(1+x)^2}$.

   b Work out the first three terms in the binomial expansion of $\dfrac{1}{(1-x)^2}$.

2 **a** Work out the coefficient of x^3 in the binomial expansion of $(1+x)^{\frac{5}{2}}$.

   b Work out the coefficient of x^4 in the binomial expansion of $(1+x)^{\frac{5}{2}}$.

3 **a** Work out the first three terms of the binomial expansion of $\sqrt[3]{1+x}$.

   b Write down the range of values of x for which the expansion is valid.

   c Use the expansion to find an approximation to $\sqrt[3]{1.1}$.
      Round your answer to 3 d.p.

4 **a** Work out the binomial expansion for $\sqrt{1+x}$ as far as the term in x^3.

   b Find the binomial expansion for $\sqrt{1+2x}$ as far as the term in x^3.

   c Find the range of values of x for which the expansion in **part b** is valid.

5 **a** Work out a cubic approximation to $\dfrac{1}{1+x}$ for small x.

   b Use your answer to **part a** to the find an approximation to $\dfrac{1}{1+x^2}$ for small x.

6 Find the binomial expansion of $(2+3x)^{-2}$ as far as the term in x^3. State the range of values of x for which your expansion is valid.

7   Find the coefficient of x^3 in the binomial expansion of $\dfrac{1}{\sqrt{1-2x}}$.

PS 8   Use a binomial approximation to find a value of $4.01^{\frac{3}{2}}$ correct to 5 decimal places. Show your method.

9   Find the binomial expansion of $\dfrac{1}{\sqrt{1-3x}}$ as far as the term in x^5.

10   Use a binomial approximation to find a value of $3.99^{\frac{1}{2}}$ correct to 5 decimal places.

SUMMARY OF KEY POINTS

› The modulus function is of the form $|x|$.

> **›** $|x| = x$ for $x \geqslant 0$
>
> **›** $|x| = -x$ for $x < 0$, i.e. all the values of $|x|$ for $x < 0$ are positive.

› Useful modulus relations:

> **›** $|a| = |b| \Leftrightarrow a^2 = b^2$
>
> **›** $|x - a| < b \Leftrightarrow a - b < x < a + b$ or $|x - a| \leqslant b \Leftrightarrow a - b \leqslant x \leqslant a + b$
>
> **›** $|x - a| \geqslant b \Leftrightarrow x \leqslant a - b$ or $x \geqslant a + b$ or $|x - a| > b \Leftrightarrow x < a - b$ or $x > a + b$.

› Algebraic division:

> **›** The expression found after the algebraic division is known as the quotient.

› Factor theorem:

> **›** $f(a) = 0 \Leftrightarrow (x - a)$ is a factor of $f(x)$.

› Remainder theorem:

> **›** If $f(a) \neq 0$, then $(x - a)$ is *not* a factor of $f(x)$ and $f(a)$ is the remainder.
>
> **›** $F(x) = Q(x) \times$ divisor + remainder.

› Partial fractions:

> **›** Partial fractions are two or more fractions into which a more complex fraction can be split.
>
> **›** An expression with two or more linear terms in the denominator, for example, $\dfrac{f(x)}{(x+p)(x+q)(x+r)}$, can be split into partial fractions in the form $\dfrac{A}{x+p} + \dfrac{B}{x+q} + \dfrac{C}{x+r}$, where A, B and C are constants.
>
> **›** An expression that has repeated linear terms in the denominator, for example $\dfrac{f(x)}{(x+p)(x+q)^2}$, can be split into partial fractions in the form $\dfrac{A}{x+p} + \dfrac{B}{x+q} + \dfrac{C}{(x+q)^2}$, where A, B and C are constants.
>
> **›** For each and every quadratic factor, in the form $cx^2 + d$, in the denominator where $c \geqslant 1$, there will be a related partial fraction in the form $\dfrac{Ax + B}{cx^2 + d}$, where A and B are constants.

› Binomial expansion:

> **›** $(1 + x)^n = 1 + nx + \dfrac{n(n-1)}{2}x^2 + \dfrac{n(n-1)(n-2)}{3!}x^3 \dots$ if n is a rational number and $|x| < 1$. This is the general binomial expansion.

EXAM-STYLE QUESTIONS

1 $f(x) = |3x + 5|$ and $g(x) = 4x$

    a Solve the equation $f(x) = g(x)$, showing all your working.

    b Solve the inequality $2f(x) > |g(x)|$.

2 The polynomial $2x^3 + x^2 + px + 12$, where p is a constant, is denoted by $f(x)$. It is given that $(x - 3)$ is a factor of $f(x)$.

    a Find the value of p.

    b When p has this value, factorise $f(x)$ completely.

3 The polynomial $p(x)$ is defined by $p(x) = x^3 + 2ax^2 + 3ax + 14$, where a is a constant.

    a Given that $(x - 7)$ is a factor of $p(x)$, find the value of a.

    b When a has this value,

        i solve the equation $p(x) = 0$

        ii find the remainder when $p(x)$ is divided by $(x - 2)$.

4 Solve the inequality $|3 - 2x| < |4x + 3|$.

PS 5 The polynomial $p(x)$ is defined by $p(x) = x^4 - 9x^2 - ax + 12$, where a is a constant.

    a Given that $(x - 3)$ is a factor of $p(x)$, find the value of a.

    b Use the factor theorem to identify a second factor of $p(x)$.

    c Using your numerical value of a, solve the equation $x^4 - 9x^2 = ax - 12$.

PS 6 **a** Solve the inequality $7 - 3x < |11x - 5|$.

    b Show that the inequality $7 - 3(y - 1)^2 < |11(y - 1)^2 - 5|$ can be rewritten as
       $4 + 6y - 3y < |11y^2 - 22y + 6|$.

    c Hence solve the inequality $8 - 2y - y^2 < |11y^2 - 22y + 6|$.

PS 7 **a** Fully factorise $4x^4 + 8x^3 - 21x^2 - 18x + 27$.

    b Solve $6x^4 + 2x^3 - 10x = 2x^4 - 6x^3 + 21x^2 + 8x - 27$.

C 8 When the expressions $mx^3 - 3x^2 + 5x - 4m$ and $3x^3 - 5x^2 - mx + 2m$ are each divided by $(x - 3)$, the remainders are the same. Find the value of m.

C 9 Find the remainder, if any, when $8x^3 - 8x^2 + 8x - 3$ is divided by $4x^2 - 2x + 3$. Give your answer in the form $(ax + b)$, where a and b are integers.

10 **a** It is given that when $x^4 - x^3 + 2x^2 + px + q$ is divided by $(x^2 + 2x + 3)$, the remainder is zero. Find the values of the constants p and q.

    b When p and q have these values, show that there are no solutions to the equation
       $x^4 - x^3 + 2x^2 + px + q = 0$.

11 a Find the quotient and remainder when $4x^3 + 20x^2 + px - 60$ is divided by $(2x^2 + 13x + 20)$.

b Given that $(2x^2 + 13x + 20)$ is a factor of $4x^3 + 20x^2 + px - 60$, write down the value of p.

c For this value of p, solve the equation $4x^3 + 20x^2 + px - 60 = 0$.

(C) 12 It is given that $f(x) = \left| 2 - \frac{1}{2}x \right|$ and $g(x) = |2x + 4|$ for $-5 < x < 5$.

a On the same set of axes, sketch the graphs of $y = f(x)$ and $y = g(x)$, stating clearly the coordinates of any points where the graphs meet the x-axis or the y-axis.

b Solve the inequality $\dfrac{|2x + 4|}{\left| 2 - \frac{1}{2}x \right|} = 0$.

13 Express $\dfrac{11x - 5}{(x - 3)(3x - 2)}$ in partial fractions.

14 Express $\dfrac{6x + 7}{(x + 1)^2}$ in partial fractions.

15 Express $\dfrac{9}{(x + 1)(x^2 + 2)}$ in partial fractions.

16 a Expand $\dfrac{1}{1 - x}$ in ascending powers of x up to and including the term in x^3, simplifying the coefficients.

(C)

b Hence, or otherwise, expand $\dfrac{1}{2 - x}$ in ascending powers of x up to and including the term in x^3, simplifying the coefficients.

c State the values of x for which the expansion is valid.

(C) 17 Let $f(x) = \dfrac{16x^2 + 29x + 7}{(x + 4)(2x + 1)^2}$.

a Express $f(x)$ in partial fractions.

b Hence obtain the expansion of $f(x)$ in ascending powers of x, up to and including the term in x^2.

(PS) 18 a Express $\dfrac{x^2 + 1}{(x - 2)^3}$ in partial fractions.

b Hence obtain the expansion of $\dfrac{x^2 + 1}{(x - 2)^3}$ in ascending powers of x, up to and including the term in x^2.

c State the values of x for which the expansion is valid.

(C) 19 a Expand $\sqrt{4 - 8x}$ in ascending powers of x up to and including the term in x^3, simplifying the coefficients.

b State the values of x for which the expansion of $\sqrt{4 - 8x}$ is valid.

(PS) 20 Given that $\dfrac{7x - 5}{(x - a)(x - 3)} \equiv -\dfrac{9}{x - a} + \dfrac{b}{x - 3}$, where a and b are constants. Find the values of a and b.

21 Let $f(x) = \dfrac{3x^2 + 4x + 1}{x^3 - x^2 + 2x - 2}$.

    a Express $f(x)$ in partial fractions.

    b State the values of x for which the expansion is valid.

Mathematics in life and work

The distance travelled by vehicle A at time t s is given by $d = 7 - t$. When $t = 4$, vehicle B starts its journey. At time t s, the distance travelled by vehicle B is given by $d = 2t - 8$.

1 On the same set of axes draw the distance–time graphs of vehicles A and B.

2 After how many seconds do vehicles A and B meet?

3 Solve the equation $|7 - t| = |2t - 8|$.

4 By comparing your answers to **parts 2** and **3**, state how many times the vehicles meet. You must justify your answer.

2 LOGARITHMS AND EXPONENTIAL FUNCTIONS

Mathematics in life and work

Sometimes when quantities change they do so in a linear way – in which case a graph showing changes over time is a straight line.

However, the change is often more complicated than that and the graph will be a curve. Finding the equation of the line helps you to understand the relationship between the variables. Logarithms and exponential functions are a useful tool in doing this.

There are a range of careers that might involve using logarithms and exponential functions – for example:

> If you were an investment fund manager, you could use exponential functions to model the growth in value of shares and make predictions about potential future value.

> If you were an engineer working in a nuclear power station, you would need to use exponential functions to assess the radioactivity of waste material when planning its safe disposal.

> If you were a seismologist measuring the strength of tremors and earthquakes on the Richter scale, you would be using logarithms to do this.

> If you were a town planner looking at future infrastructure needs in your area, you would use exponential functions to model future population changes.

LEARNING OBJECTIVES

You will learn how to:

> understand the relationship between logarithms and indices and use the laws of logarithms

> understand the definition and properties of e^x and $\ln x$, including their relationship as inverse functions, and their graphs

> use logarithms to solve equations of the form $a^x = b$, and similar inequalities

> use logarithms to transform a given relationship to linear form, and hence determine unknown constants by considering the gradient and/or intercept.

LANGUAGE OF MATHEMATICS

Key words and phrases you will meet in this chapter:

> base, index, logarithm, natural logarithm, parameter

PREREQUISITE KNOWLEDGE

You should already know how to:

» manipulate expressions involving powers of numbers

» use the equation of a graph to draw it

» work out the equation of a straight line through given points

» solve linear equations.

You should be able to complete the following questions correctly:

1 Write as a power of 2:

    a 16   **b** $\frac{1}{32}$ **c** $\sqrt{2}$

    d $\sqrt{8}$ **e** 1.

2 Write as a power of n:

    a $n^2 \times n^3$ **b** $n^2 \div n^3$

    c $(n^{-3})^{-2}$ **d** $\sqrt{n} \times \sqrt[3]{n}$.

3 Sketch these graphs for $x > 0$:

    a $y = 10 + 0.5x$ **b** $y = 0.5x^2$ **c** $y = 2x^{0.5}$.

4 Work out the equation of a straight line that goes through (5, 18) and crosses the y-axis at (0, 20).

5 The price of a car was $24 000. The price is increased by 4%, then it is increased again by 4%.

    Work out the new price.

2.1 Logarithms

Here are some powers of 10:

» $100 = 10^2$

» $10\,000 = 10^4$

» $0.001 = 10^{-3}$

The **logarithm** of a number is the power of 10 that is equal to that number. So the logarithm of 100 is 2, the logarithm of 10 000 is 4 and the logarithm of 0.001 is –3.

'log' is the abbreviation for logarithm. So, you would write:

$\log 1\,000\,000 = 6$ and $\log 0.000\,01 = -5$.

Every positive number has a logarithm. For example,

$10^{\frac{1}{2}} = \sqrt{10} = 3.16228$ to 5 d.p.

So $\log 3.16228 = 0.5$.

Stop and think	Why is it possible to find a logarithm only for positive numbers and not for zero or negative numbers?

Raising 10 to a power and finding the logarithm are inverse operations.

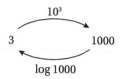

You already know these laws of indices:

1 $10^a \times 10^b = 10^{a+b}$ To multiply powers, add the indices.

2 $10^a \div 10^b = 10^{a-b}$ To divide powers, subtract the indices.

3 $(10^a)^c = 10^{ac}$ To find a power of a power, multiply the indices.

Corresponding to each of these, there are three laws of logarithms:

1 $\log xy = \log x + \log y$

2 $\log \dfrac{x}{y} = \log x - \log y$

3 $\log x^k = k \log x$

To show that these are true, suppose $x = 10^a$ and $y = 10^b$:

❯ $xy = 10^a \times 10^b = 10^{a+b}$

Hence $\log xy = a + b = \log x + \log y$.

❯ $\dfrac{x}{y} = 10^a \div 10^b = 10^{a-b}$

Hence $\log \dfrac{x}{y} = a - b = \log x - \log y$.

❯ $x^k = (10^a)^k = 10^{ak}$

Hence $\log x^k = ka = k \log x$.

> **KEY INFORMATION**
>
> You need to learn the three laws of logarithms:
>
> $\log xy = \log x + \log y$
>
> $\log \dfrac{x}{y} = \log x - \log y$
>
> $\log x^k = k \log x$

Example 1

$\log x = c$

Write, in terms of c, the logarithms of:

a x^3 **b** $\dfrac{10}{x}$ **c** $100\sqrt{x}$.

Solution

a $\log x^3 = 3 \log x = 3c$

b $\log \dfrac{10}{x} = \log 10 - \log x$

$= 1 - c$

c $\log 100\sqrt{x} = \log 100 + \log \sqrt{x}$

$$= 2 + \log x^{\frac{1}{2}}$$

$$= 2 + \frac{1}{2}\log x$$

$$= 2 + \frac{1}{2}c$$

Stop and think Identify where each of the three laws of logarithms has been used in **Example 1**. Explain why $\log 10 = 1$ and $\log 1 = 0$.

Calculators and spreadsheets usually have a logarithm function. Make sure you know how to use this. Check that you can correctly find the logarithms of some numbers such as 1000 or 0.001 or $\sqrt[3]{10}$.

Make sure that you can find the logarithm of any non-negative number.

Exercise 2.1A

1 Without using a calculator, work out the value of:

a $\log 1000$

b $\log 1\,000\,000$

c $\log 0.001$

d $\log 0.000\,001$.

2 Without using a calculator, work out the value of:

a $\log \sqrt{10}$

b $\log \sqrt{100}$

c $\log \sqrt{1000}$

d $\log \dfrac{1}{\sqrt{10}}$.

3 $\log 2 = 0.3010$ to 4 d.p.

Without using a calculator, use this fact to work out the value of:

a $\log 20$

b $\log 0.2$

c $\log 5$

d $\log \dfrac{1}{\sqrt{2}}$.

4 $\log 8 = 0.9031$ to 4 d.p. and $\log 12 = 1.0792$ to 4 d.p.

Without using a calculator, use this fact to work out the value of:

a $\log 80$

b $\log 1.5$

c $\log 9.6$

d $\log\left(\dfrac{2}{3}\right)$.

5 $\log x = k$

Write the following in terms of k.

a $\log x^2$

b $\log 100x$

c $\log \dfrac{1}{x}$

d $\log \dfrac{10}{\sqrt{x}}$.

6 $\log x = k$ and $\log y = h$

Write the following in terms of k and h.

a $\log xy^2$

b $\log \dfrac{\sqrt{x}}{y}$

c $\log \left(\dfrac{100x^2}{y^3} \right)$

d $\log \dfrac{1}{xy}$.

7 Do not use a calculator for this question.

a $\log x = 2$. Work out the value of x.

b $\log 2y = 2$. Work out the value of y.

c $\log (\log z) = 2$. Work out the value of z.

PS **8** $\log 2 = 0.3010$ and $\log 3 = 0.4771$.

You can use these facts to work out the logarithms of other numbers without using a calculator, but just using simple arithmetic.

Which of the following can be found easily in this way, without using a calculator?

a $\log 4$

b $\log 5$

c $\log 6$

d $\log 7$

e $\log 8$

f $\log 9$

g $\log 11$

h $\log 12$

Show how to do so.

9 **a** a is a positive number and k is a rational number.

Prove, using the definition of a logarithm, that $\log a^k = k \log a$.

b Prove that $\dfrac{\log 216}{\log 36} = 1.5$.

10 **a** Prove that $\log \dfrac{1}{a} = -\log a$.

b Prove that if n is a positive integer, then $\log a^n = n^2 \log \sqrt[n]{a}$.

11 $f(x) = \log x$, $x > 0$

P and Q are two points on the graph of $y = f(x)$ with x-coordinates a and $a + 1$.

a Find the gradient of the chord PQ.

b Show that the function is an increasing function.

c Find the limit of the gradient of PQ as a increases.

12 **a** Sketch a graph of $y = 10^x$.

b Explain why a graph of $y = \log x$ is a reflection in the line $y = x$.

c It can be shown that if $y = 10^x$, then $\frac{dy}{dx} = 2.3 \times 10^x$.

Use this fact to find the point on the graph of $y = \log x$ where the gradient is 0.2.

2.2 Logarithms in other bases

Look at the powers of 2:

$2^5 = 32$

$2^{-3} = \frac{1}{8}$

$2^{0.5} = 1.414\,21$

You can use logarithms to write these expressions like this:

$\log_2 32 = 5$

$\log_2 \frac{1}{8} = -3$

$\log_2 1.414\,21 = 0.5$

In these cases, the **base** of the logarithm is 2. You read $\log_2 32$ as 'log to the base 2 of 32'. The subscript 2 means that you are using powers of 2 in this case.

In **Section 2.1,** all the logarithms were to base 10.

$\log 32 = 1.505\,15$; if there is no subscript assume that the logarithm is to base 10, so it is a power of 10.

$\log_2 32 = 5$; the subscript means you need a power of 2.

If necessary, when you are working with logarithms in several bases, you can write in the base 10: $\log_{10} 32 = 1.505\,15$.

You can choose any positive integer as the base of a logarithm.

Here are some equivalent statements:

$5^3 = 125 \Leftrightarrow \log_5 125 = 3$

$8^{\frac{1}{3}} = 2 \Leftrightarrow \log_8 2 = \frac{1}{3}$

$\frac{1}{49} = 7^{-2} \Leftrightarrow \log_7 \frac{1}{49} = -2$

The three laws of logarithms that you saw above for logarithms in base 10 are true for logarithms in any base.

> **KEY INFORMATION**
>
> If $y = a^x$, then $\log_a y = x$. The base of the logarithm is a.
>
> ❯ $\log_a xy = \log_a x + \log_a y$
>
> ❯ $\log_a \frac{x}{y} = \log_a x - \log_a y$
>
> ❯ $\log_a x^k = k \log_a x$

Stop and think Look through the explanations of the three laws of logarithms (base 10) given earlier in **Section 2.1**. How can you change them so that they are valid for logarithms to any base?

Example 2

Find: **a** $\log_3 243$ **b** $\log_4 32$.

Solution

a You have to write 243 as a power of 3.

$243 = 3^5 \Leftrightarrow \log_3 243 = 5$

b $32 = 16 \times 2$

$= 4^2 \times \sqrt{4}$

$= 4^2 \times 4^{\frac{1}{2}}$

$= 4^{2.5}$

$\Rightarrow \log_4 32 = 2.5$

An alternative way of thinking about this question is to use the result that $\log_a a = 1$.

$\log_3 243 = \log_3 3^5$

$= 5 \log_3 3$

$= 5$

You may have a $\log_a b$ button on your calculator. If so, use it to check the values in Example 2 using the $\log_a b$ button.

However, it is useful to be able to recognise the answer in simple cases when the answer is a small integer.

Exercise 2.2A

1 Without using a calculator, work out the value of:

   a $\log_2 64$ **b** $\log_5 125$

   c $\log_3 243$ **d** $\log_4 256$.

2 Without using a calculator, work out the value of:

   a $\log_9 3$ **b** $\log_9 27$

   c $\log_9 \dfrac{1}{81}$ **d** $\log_9 \sqrt{3}$.

3 Match each expression on the left to a corresponding one on the right.
   One has been done for you.

   $\log_a 2 + \log_a 3$ $\log_a 3$

   $2 \log_a 4$ $\log_a 4$

   $\log_a 8 - \log_a 2$ $\log_a 6$

   $\log_a \dfrac{1}{2} - \log_a \dfrac{1}{6}$ $\log_a 8$

   $3 \log_a 2$ $\log_a 16$

4 Do not use a calculator for this question.

Here is a table of values of powers of 2.

x	1	2	3	4	5	6	7	8	9	10
2^x	2	4	8	16	32	64	128	256	512	1024

Use the table to find:

a $\log_2 128$ **b** $\log_2 \frac{1}{32}$

c $\log_2 \sqrt{512}$ **d** $\log_2 \sqrt[3]{256}$.

5 $\log_3 5 = c$

Without using a calculator, write the following in terms of c.

a $\log_3 15$ **b** $\log_3 0.2$

c $\log_3 125$ **d** $\log_9 5$

6 Find the base for the following logarithms.

a $\log_n 216 = 3$ **b** $\log_n 4 = \frac{1}{2}$

c $\log_n 1 = 0$ **d** $\log_n 27 = 1.5$

(PS) 7 $\log_n a = 3.5$ and $\log_n b = 4.5$

Prove that $b = na$.

(PS) 8 **a** Write each of these equations in index form.

     i $x = \log_a b$ **ii** $y = \log_b a$

   b Show that $\log_a b = \dfrac{1}{\log_b a}$.

9 **a** Show that $\dfrac{\log_2 16}{\log_2 8} = \log_8 16$.

   b Prove that $\log_b a = \dfrac{\log_c a}{\log_c b}$.

10 Solve the following equations.

   a $\log_{x-1} 1024 = 5$

   b $\log_x 9x = 3$

11 Solve the following simultaneous equations.

$\log_4 x + \log_4 y = 6$

$\log_4 x - \log_4 y = 3$

12 **a** Show that $\log_{a^2} x = \frac{1}{2} \log_a x$.

   b Solve the equation $\log_2 x + \log_4 x = 12$.

   c Solve the equation $\log_2 x + \log_4 x + \log_8 x = 22$.

2.3 The number e

Here is the graph of $y = 2^x$.

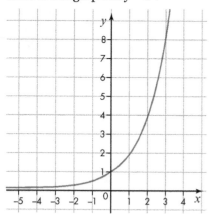

How can you find the gradient at a particular point with coordinates $(x, 2^x)$?

You can draw the tangent to the curve at any point. The gradient of the tangent at that point is the gradient of the curve at that point.

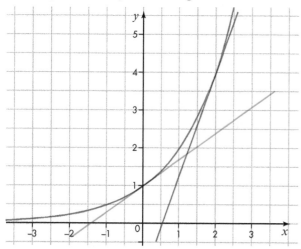

This graph has tangents drawn at (0, 1) and (2, 4).

By drawing a triangle under the tangent, you can calculate the gradient.

Here is a table to show the gradient at different points on the curve $y = 2^x$

Point	(−3, 0.125)	(−2, 0.25)	(−1, 0.5)	(0, 1)	(1, 2)	(2, 4)	(3, 8)
Gradient	0.087	0.173	0.347	0.693	1.386	2.773	5.545

The gradients are rounded to 3 decimal places.

If you look carefully, you will see that the gradient is always $0.6931 \times$ the y-coordinate.

> The value of the multiplier is given to 4 d.p.

This is true for any point on the curve, not just for those in the table.

The gradient of the curve $y = 2^x$ at (x, y) is $0.6931y$.

This method could be used for any curve of the form $y = a^x$ where a is a positive number.

In every case, the gradient is a multiple of the y-coordinate.

Here are some expressions for the gradient for different values of a.

y	2^x	3^x	5^x	1.5^x	0.8^x
Gradient	$0.6931y$	$1.0986y$	$1.6094y$	$0.4055y$	$-0.2231y$

Stop and think Why does the multiplier for $y = 0.8^x$ have a minus sign?

$y = a^x$, where $a > 0 \Rightarrow$ the gradient is always a multiplier $\times y$.

The larger a is, the larger the multiplier is.

If $a < 1$, then the multiplier is negative.

From the table, it looks as if there should be a value of a between 2 and 3 for which the multiplier is 1.

There is! It is an irrational number (like π) with an infinite decimal expansion that does not recur.

It is denoted by the letter e. Here are the first few digits of the decimal expansion of e.

e = 2.718 281 828 459 045 ...

We now have this important result:

$y = e^x \Rightarrow$ the gradient at (x, y) is e^x. You can write this as

$$\frac{dy}{dx} = e^x = y.$$

The gradient at any point on the curve is equal to the y-coordinate of that point.

Your calculator should have a key for finding powers of e.

Use it to confirm the value of e shown above.

Here is a graph of $y = e^x$.

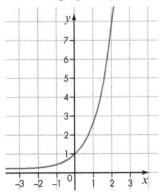

Check by eye that the gradient where $y = 1, 2, 3$ or $\frac{1}{2}$ is indeed 1, 2, 3 or $\frac{1}{2}$.

KEY INFORMATION

The gradient at a point on the curve $y = e^x$ is the y-coordinate, or e^x.

$$y = e^x \Rightarrow \frac{dy}{dx} = e^x = y.$$

For comparison, here are graphs of $y = 2^x$, $y = e^x$ and $y = 3^x$.

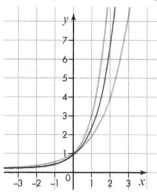

Now look more closely at the graph of $y = e^{2x}$.

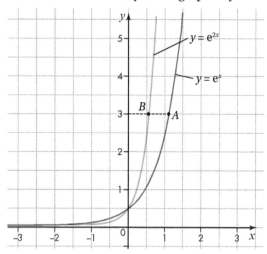

The graph of $y = e^{2x}$ is a stretch of $y = e^x$ from the y-axis with a scale factor of $\frac{1}{2}$.

This means that gradients at corresponding points are multiplied by 2. For example, the gradient at A is 3 and the gradient at B is 6.

The gradient at B is $2 \times$ the y-coordinate and this is true for any point on the curve $y = e^{2x}$.

$y = e^{2x} \Rightarrow$ the gradient $= 2e^{2x} = 2y$.

The generalisation is true: $y = e^{ax} \Rightarrow$ the gradient $= ae^{ax} = ay$ for any value of a.

A graph of $y = e^{ax}$ has the following properties:

> it crosses the y-axis at $(0, 1)$

> it is always above the x-axis

> if $a > 0$, the gradient is always positive and increases as x increases

> if $a < 0$, the gradient is always negative and gets closer to 0 as x increases.

KEY INFORMATION

$y = e^{ax} \Rightarrow$ gradient

$\dfrac{dy}{dx} = ae^{ax} = ay$ for any value of a.

Example 3

a Sketch the graph of $y = e^{-0.5x}$.

b Find the gradient where the curve crosses the y-axis.

c Find the y-coordinate of the point on the graph where the gradient is -0.1.

Solution

a The graph is similar in shape to $y = e^{-x}$, getting less steep and closer to the x-axis as x increases.

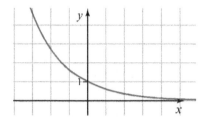

b $y = e^{-0.5x} \Rightarrow$ the gradient is $-0.5e^{-0.5x}$.

Where the graph crosses the y-axis, $x = 0$ and the gradient is $-0.5 \times e^0 = -0.5$.

c If the gradient is -0.1, then $-0.5y = -0.1$ and hence $y = 0.2$.

Exercise 2.3A

1 Find the value of these expressions to 3 s.f.

    a e^2 **b** $\sqrt{e}$ **c** $\dfrac{1}{e}$

    d $\dfrac{e+2}{e-2}$ **e** e^{-3}

2 **a** Find the value of $\left(1 + \dfrac{1}{n}\right)^n$ when:

    i $n = 100$ **ii** $n = 10\,000$ **iii** $n = 1\,000\,000$.

    b How are the answers to **part a** connected to the value of e?

    c What do your answers suggest about the limit of $\left(1 + \dfrac{1}{n}\right)^n$ as $n \to \infty$?

3 **a** On the same axes, sketch the curves:

    i $y = e^x$ **ii** $y = e^{-x}$.

    b Describe the transformation that will map one curve onto the other.

4 The equation of a curve is $y = e^x$. Find the gradient of the curve:

    a at $(0, 1)$

    b at the point with an x-coordinate of 2

    c at the point with a y-coordinate of 2.

5 The equations of these curves are $y = e^x + 1$, $y = e^{x+1}$ and $y = (e + 1)^x$.

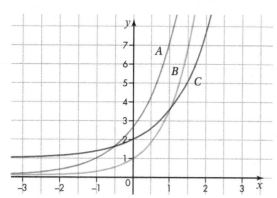

a Match each equation to the appropriate curve on the graph.

b Two of the curves are translations of $y = e^x$. Identify the curves and give the vector of the translation in each case.

c One of the curves is a stretch of $y = e^x$. Identify the curve and describe the stretch.

6 **a** Find the point where the curve $y = e^{x+2}$ crosses the y-axis.

b Show that the curves $y = e^{x+2}$ and $y = e^{x+4}$ have exactly the same shape.

7 **a** Sketch the graphs of $y = e^{0.5x}$ and $y = e^{2x}$ on the same axes.

b Describe a stretch that will map $y = e^{2x}$ onto $y = e^{0.5x}$.

8 Find the exact solutions of the equation $2x^2 + ex = e^2$.

9 The equation of a curve is $y = e^{0.5x+a}$.

a Show that $2\dfrac{dy}{dx} = y$.

b The curve crosses the y-axis at A. Find the coordinates of A.

c Show that the tangent at A crosses the x-axis at $(-2, 0)$.

10 $f(x) = e^x$

The point P is on the graph of $y = f(x)$.

The tangent at P passes through the origin.

a Find the coordinates of P.

b Show that the equation of the normal at P can be written as $x + ey = 1 + e^2$.

11 $f(x) = 4e^{2x+1}$

a Show that $f'(x) = 2f(x)$.

b P is the point where the graph of $y = f(x)$ crosses the line $x = -1$.

Find the coordinates of P.

c Find the point where the tangent at P crosses the x-axis.

12 *A* is the point (a, e^a) on the graph $y = e^x$.

The tangent and normal to the curve at *A* meet the *x*-axis at *T* and *N*, respectively.

Find the area of triangle *ATN*.

2.4 Natural logarithms

In **Section 2.3**, you looked at powers of e. In this section, you will look at logarithms to the base e.

$e^2 = 7.389$ to 4 s.f.

You could write that as $\log_e 7.389 = 2$.

Logarithms to the base e are sometimes called **natural logarithms** and they are written as ln.

So, you write $\ln 7.389 = 2$.

Here are some values of $\ln x$, given to 3 d.p.

x	10	100	1000	0.1	0.01
$\ln x$	2.303	4.605	6.908	−2.303	−4.605

Here is a graph of $y = \ln x$.

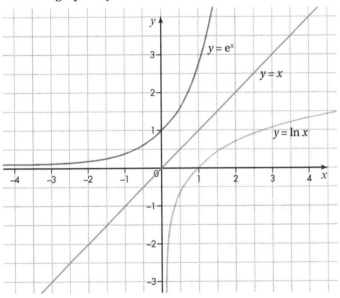

The graph of $y = \ln x$ is a reflection of the graph of $y = e^x$ in the line $y = x$. This is because e^x and $\ln x$ are inverse functions.

Stop and think   Describe the properties of the graph of $y = \ln x$.

> **KEY INFORMATION**
>
> $\log y$ means logarithm to the base 10 and $\ln y$ means logarithm to the base e, where $e \approx 2.71828$
>
> $e^x = y \Leftrightarrow \ln y = x$.

Your calculator should have a key for finding ln. It is often the same key as e^x because these are inverse functions.

Example 4

Solve the equation $20e^{2x-4} = 35$.

Solution

Divide both sides of the equation by 20.

$e^{2x-4} = 1.75$

Rewrite this in logarithm form.

$2x - 4 = \ln 1.75$

Rearrange.

$x = \dfrac{\ln 1.75 + 4}{2} = 2.280$ to 4 s.f.

> This step is sometimes known as 'taking the log of both sides'.

Exercise 2.4A

1 $\ln a = 3$ and $\ln b = 4.5$

Work out the value of each expression.

a $\ln ab$

b $\ln a^4$

c $\ln \dfrac{b}{a^2}$

d $\ln \sqrt{ab}$

2 Solve these equations. Give your answers to 3 s.f.

a $e^x = 2000$

b $e^{-x} = 0.03$

c $e^{2x-5} = 125$

d $e^{-\frac{1}{2}x^2} = 0.5$

3 Solve these equations. Give your answers to 3 s.f.

a $\ln 0.5x = 4$

b $\ln(4x + 2) = 3.5$

c $4 + \ln 2x = 0$

d $\ln \dfrac{120}{x} = 5$

4 What is the connection between $\ln a$ and $\ln \dfrac{1}{a}$ for $a > 0$?

5 **a** Show that the graph of $y = \ln 4x$ is a translation of the graph of $y = \ln x$.

Write down the vector that describes the translation.

b On the same axes, sketch the graphs of $y = \ln x$ and $y = \ln 4x$.

6 Solve these equations. Give your answers to 3 d.p.

a $20e^t = 100$

b $40e^{-t} = 35$

c $250e^{3t} = 8000$

d $32.5e^{-0.85t} = 14.8$

7 **a** Draw a sketch to show that the graphs of $y = 2e^x$ and $y = 6e^{-x}$ cross at one point.

b Find the coordinates of the point of intersection.

8 The graph of $y = \ln x$ is a reflection
of the graph of $y = e^x$ in the line $y = x$.

Point A has the coordinates $(2, \ln 2)$
and B is the reflection of A.

a Write down the coordinates of B.

b Find the gradient of $y = e^x$ at B.

c Find the gradient of $y = \ln x$ at A.

d Generalise your result from
part c.

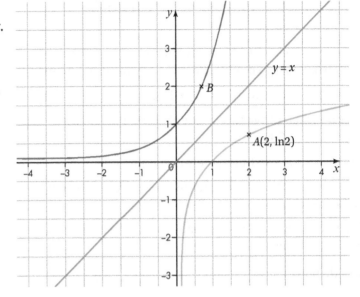

9 Here is an equation: $e^{2x} + e^x = 6$.

a Make the substitution $y = e^x$ and show that you get a quadratic equation in y.

b Solve the quadratic equation to get two values for y.

c Use your values for y to solve the original equation.

10 Solve the equation $2e^{2x} - 9e^x + 4 = 0$.

11 Show that:

a $\ln (e + e^2) \equiv 1 + \ln (1 + e)$

b $\ln (e^2 - e^4) \equiv 2 + \ln (1 + e) + \ln (1 - e)$.

12 $f(x) = 2e^x + e^{-x}$

a Show that the graph of $y = f(x)$ has a stationary point and find its coordinates.

b Show that $f(x)$ has a minimum value of $2\sqrt{2}$.

13 a Show that $y = 10^x$ can be written as $y = e^{cx}$ for a particular value of c and find this value.

b If $y = 10^x$, find $\dfrac{dy}{dx}$.

c Find the x-coordinate of the point where the gradient is 10.

14 The equation of a curve is $y = 0.2e^{0.5x}$.

The curve crosses the line $y = a$ at Q.

a Find the x-coordinate of Q.

The tangent at Q meets the x-axis at $(6, 0)$.

b Find the value of a.

2.5 Using logarithms to solve equations and inequalities

Suppose you invest \$400 at an interest rate of 6% per year. How long will it take to double the value of your investment?

The value after n years is $\$400 \times 1.06^n$.

To find when the value has doubled you need to solve the equation $400 \times 1.06^n = 800$.

Divide by 400 and the equation becomes $1.06^n = 2$.

How do you solve this equation? The answer is to take logarithms of both sides.

$\log 1.06^n = \log 2$ and hence, using one of the laws of logarithms, $n \log 1.06 = \log 2$.

Divide both sides by $\log 1.06$ to get $n = \dfrac{\log 2}{\log 1.06} = 11.90$.

In the context of the original question, the answer needs to be a whole number of years.

It takes 12 years before the value of the investment has doubled.

> You do not need to find the actual values of the logarithms. Just use your calculator to do the final division.

> You could have used logarithms to any base to rewrite the equation, but it is best to use base 10 or base e because you can easily find logarithms to base 10 or base e on your calculator.
>
> If you use natural logarithms to solve that equation, you get $n = \dfrac{\ln 2}{\ln 1.06}$.
>
> Check that this gives the same answer.

Example 5

Solve the equation $2^{4x-1} = 7^{x+1}$.

Solution

Take logarithms of both sides.

$$\log 2^{4x-1} = \log 7^{x+1}$$

$$\Rightarrow (4x - 1)\log 2 = (x + 1)\log 7$$

Rearrange.

$$4x \log 2 - x \log 7 = \log 7 + \log 2$$

$$\Rightarrow x = \frac{\log 7 + \log 2}{4 \log 2 - \log 7} = 3.192 \text{ to 4 d.p.}$$

> You can use logarithms to base 10 or base e here. Both give the same answer.

A similar method can be used to solve inequalities involving indices.

The logarithm is an increasing function.

$a < b \Leftrightarrow \ln a < \ln b$, and $a > b \Leftrightarrow \ln a > \ln b$.

This means that you can take logarithms in an inequality and the inequality does not change.

Example 6

Solve the inequality $5^{3x-1} < 12$.

Find the value of x to 3 d.p.

Solution

Take logarithms of both sides.

$\ln 5^{3x-1} < \ln 12$ 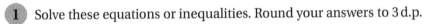 The inequality is still valid.

$(3x - 1) \ln 5 < \ln 12$

$3x - 1 < \dfrac{\ln 12}{\ln 5} = 1.5439\ldots$

$x < \dfrac{1.5439 + 1}{3} \Rightarrow x < 0.848$ to 3 d.p.

Exercise 2.5A

1 Solve these equations or inequalities. Round your answers to 3 d.p.

   a $3^x = 11$ **b** $4^x = 175$ **c** $12^x < 6$

2 Solve these equations or inequalities.

   a $0.5^x = 0.4$ **b** $0.7^x \geqslant 0.25$ **c** $0.9^x = 2.55$

3 Solve these equations.

   a $200 \times 1.8^t = 750$ **b** $7000 \times 0.87^t \leqslant 4500$ **c** $95 \times 1.04^t > 123$

4 Sadiq has some shares. Their current value is \$12 500. The value of the shares is increasing by 9% a year.

   If this rate does not change, work out how long it will take for the value of the shares to:

   a increase by 50%   **b** double.

5 Solve these equations.

   a $4^{x+2} = 90$ **b** $6^{2x+1} > 35$ **c** $15^{4x-3} = 8$

C **6** There is 10 m² of weed on a pond. The amount of weed is increasing by 50% each week.

   a Find a formula for the area (y m²) of weed on the pond in t weeks' time.

   b Find the area of weed in three weeks' time.

   c The area of the whole pond is 80 m². How long will it take until the whole pond is covered with weed?

   d Increasing the flow of water through the pond decreases the rate of growth to 20% per week. How long would it take now to cover the pond?

   e Explain why either model might not be appropriate for a large number of weeks.

7 The number of people flying from an airport this year was 1.65 million. The number of travellers is expected to increase by 5% per year.

   a How many travellers are expected next year and the year after?

   b Write down an expression for the annual number of travellers in t years' time.

   c How long will it take until the annual number exceeds 2.5 million?

   d Explain why the model might not be correct after a large number of years.

8 **a** Solve the equation $5^{x+3} = 7^{x-1}$.

   b Solve the inequality $3^{x+2} > 4^x$.

9 $a^{x+1} = (a+1)^x$ where a is a positive integer.

Show that $x = \dfrac{\ln a}{\ln\left(1 + \dfrac{1}{a}\right)}$.

10 Moore's law in computing says that the number of transistors in an integrated circuit doubles approximately every two years. This is because improvements in technology make it possible to make the transistors smaller and smaller.

   a According to Moore's law, how many years will it take for the number of transistors in an integrated circuit to increase by a factor of 100?

   b Sometimes it is said that the time for the number of transistors to double is only 1.5 years. How does this alter your answer to part **a**?

   c Some experts believe that Moore's law will no longer apply in the near future. Can you think of a reason for this?

11 **a** Solve the equation $2^{x^2} = 3$.

   b Solve the equation $2^{x^2} = 3^{x+5}$.

12 A geometric progression starts: 60   51   43.35   ...

   a Find the sum to infinity.

   b How many terms must be added before the total is more than 390?

13 The value of an apartment is $120 000.

The value is expected to increase by 4% per year.

   a Find the value in three years' time.

   b Find the number of years until the value is $200 000.

14 The value of a new car decreases by 15% in the first year and then by 5% a year after that.

A new car costs $22 000.

   a Find the value after 2 years.

   b Find how long it takes for the value to be halved.

15 Sam buys two tickets in a prize draw each month.

There are 100 tickets and one prize each month.

a Show that the probability that Sam does not win a prize in the next three months is approximately 0.9412.

b For how many months must Sam play so that the probability that he does not win is less than 0.5?

16 The first term of a geometric progression is 7 and the common ratio is 1.3.

a The nth term is approximately 1000. Find the value of n.

b The sum of the first k terms is approximately 10 000. Find the value of k.

Mathematics in life and work: Group discussion

You work in the planning department of a city. The current population is 350 000.

New building and demographic factors mean that the population is expected to grow by 3% a year for the next 10 years.

1 Can you predict the population in 10 years' time?

The schools for children aged from 6 to 11 in the city are currently full. More schools need to be built to deal with the increase in population.

2 Estimate the number of new school places that will be needed in 10 years' time. Justify your estimate.

3 Estimate the number of new schools that need to be built each year for the next 10 years. Justify your estimate.

2.6 Logarithmic graphs

You know how to find the coordinates of points on a line if you know the equation of the line.

This section deals with the inverse question. That is, if you know some points on the line, can you find the equation? For a straight line it is relatively easy.

> **Stop and think**　How do you find the equation of a straight line if you know the coordinates of some points on the line?

For a curve it is more difficult. However, you will usually have some idea of the type of curve you are looking for. Here are two common types.

Curve type 1: $y = ax^n$

In an experiment, the length of a pendulum (L metres) and the time (T seconds) it takes to make 10 swings are recorded. This is done for several different lengths of pendulum.

Here are the results in a table.

Length (L m)	Time (T s)
0.6	15.5
0.9	19.0
1.2	21.9
1.5	24.5
1.8	26.8
2.1	29.0

These points are not in a straight line.

As a scientist, you think that an equation of the form $T = aL^n$ will fit these values with a suitable choice of the **parameters** a and n.

When you are trying to fit a model to data, the constants are usually called parameters.

Start with $T = aL^n$.

Take logarithms of both sides.

$$\log T = \log aL^n$$
$$= \log a + \log L^n$$
$$= \log a + n \log L$$

The laws of logarithms have been used to rearrange the right-hand side of the equation.

Look at the final equation. It is in the form $y = c + mx$, with $\log T$ instead of y, $\log L$ instead of x, $\log a$ instead of c and n instead of m.

If the equation is correct, a graph of $\log T$ against $\log L$ will be a straight line. The gradient will be n and the intercept on the $\log T$ axis will be $\log a$.

Here are the values of $\log T$ and $\log L$.

Length (L m)	Time (T s)	$\log L$	$\log T$
0.6	15.5	−0.22	1.19
0.9	19.0	−0.05	1.28
1.2	21.9	0.08	1.34
1.5	24.5	0.18	1.39
1.8	26.8	0.26	1.43
2.1	29.0	0.32	1.46

Here is the graph.

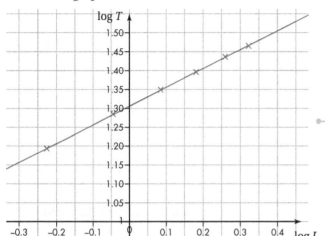

In a real experiment, there are often errors in measurements and the points will not be exactly in a straight line. In that case, you would use the line of best fit.

The points have been joined with a straight line.

You know that the equation of the straight line is $\log T = \log a + n \log L$.

The gradient is n and the intercept on the y-axis is $\log a$.

Drawing a triangle underneath the line will show that the gradient is $\frac{1}{2}$, so $n = 0.5$.

The intercept on the $\log T$ axis is 1.30, so $\log a = 1.30$.

Therefore $a = 10^{1.30} = 20.0$ to 3 s.f.

This means that the equation connecting T and L is $T = 20.0\sqrt{L}$.

Curve type 2: $y = kb^x$

If you have exponential growth or decay, the variable will be the **index** of a constant value.

A scientist is growing some bacteria in an experiment. The colony is growing.

The scientist thinks that the size of the population, P, after t hours will be given by an equation of the form $P = kb^t$.

The problem is to find the values of the parameters k and b.

Start with $P = kb^t$ and take logarithms of both sides.

$\ln P = \ln kb^t$

$\ln P = \ln k + \ln b^t$

$\quad = \ln k + t \ln b$

This is exactly the same rearrangement as in the previous type of equation but this time the variable (t) is in a different place. The graph will use values of t and not $\log t$.

t	0	2	4	6	8
P	2.4×10^3	5.6×10^4	1.3×10^6	3.0×10^7	6.9×10^8
$\ln P$	7.8	10.9	14.1	17.1	20.0

Use the values of P and t to draw a graph of $\ln P$ against t.

The gradient is 1.5.

$\ln b = 1.5$

$b = e^{1.5} = 4.48$ to 3 s.f.

The intercept on the $\ln P$ axis is 7.8.

$\ln k = 7.8$

$k = e^{7.8} = 2440$ to 3 s.f.

The equation, including these parameters, is $P = 2440 \times 4.48^t$.

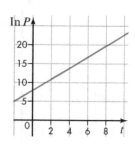

You can use logarithms to any base. In this example it is base e.

Exercise 2.6A

1 $y = 250x^2$

    a Write a formula for $\log y$ in terms of $\log x$.

    A graph of $\log y$ against $\log x$ will be a straight line.

    b Find the gradient of the line.

    c Find the coordinates of the intercept on the $\log y$ axis.

2 The formula for the volume of a sphere is $V = \frac{4}{3}\pi r^3$.
Show that a graph of $\log V$ against $\log r$ will be a straight line.

3 In physics, the gravitational force between two masses, m_1 and m_2, distance r apart, is given
by the formula $F = \dfrac{Gm_1m_2}{r^2}$, where G is a constant.

    a Show that a graph of $\ln F$ against $\ln r$ is a straight line.

    b Find the gradient of the straight line.

    c Find the intercept of the straight line on the $\ln F$ axis in terms of the constants m_1, m_2 and G.

4 The population of a country, P millions, in t years' time is modelled by the formula $P = Ac^t$,
where A and c are constants.

    a Show that a graph of $\log P$ against t is a straight line.

    The gradient of the straight line is 0.0128 and the intercept
on the $\log P$ axis is (0, 1.97).

    b Find the values of A and c.

    c Find the annual rate of growth of the population.

    d Why might the model not be accurate for a large value of t?

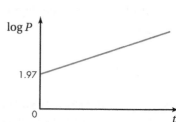

5 The formula connecting two variables, v and r, is $v = ar^n$, where a and n are constants.
Here is a graph of $\log v$ against $\log r$.

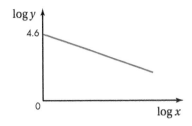

a Find the gradient of the graph.

b Explain why $\log a = 1.5$.

c Find the values of a and n.

d Use the formula to find the value of v when $r = 100$.

6 A scientist suspects that two variables are connected by the formula $x^2 y = k$, where k is a constant. The scientist collects some values for x and y and draws a graph of $\log y$ against $\log x$. The points are approximately in a straight line.

a The scientist says that this shows that the formula is valid. What is the gradient of the line?

b Estimate the value of k to 2 s.f.

7 The value of a car decreases with age, as shown in this table.

Age (t years)	2	4	6	8	10
Value ($\$y$)	18490	13675	10114	7480	5533

A model for the value of the car is the formula $y = Ac^t$, where A and c are parameters.

a Draw a suitable straight-line graph to show that this is a valid model.

b Find the values of A and c.

8 A graph of $\log y$ against $\log x$ is a straight line.

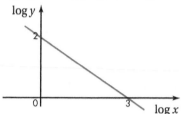

Find a formula for y in terms of x.

9 The variables p and v are connected by the formula $pv^c = k$, where c and k are constants.

Two pairs of values for p and v are given in this table.

v	25	16
p	41	80

a Show that a graph of $\log v$ against $\log p$ will be a straight line.

b Find the value of c.

c Find the value of k.

Give your answers correct to 3 s.f.

10 The variables x and t are connected by the formula $x = ka^t$, where k and a are constants.

When $t = 2$, $x = 227$ and when $t = 5$, $x = 165$.

a Find the value of a.

b Find the value of k.

c Find the value of t when $x = 100$.

11 $y \propto x^n$ where n is a constant.

Two points on a graph of y against x are (10, 8.98) and (20, 11.05).

a Explain why a graph of $\log y$ against $\log x$ will be a straight line.

b Find the gradient of the straight line in **part a**.

c Find a formula for y in terms of x.

12

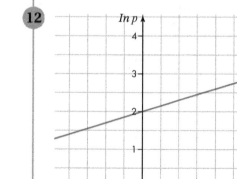

The variables p and t are linked by the formula $p = ak^t$, where a and k are constants.

a Use the graph to find the value of a.

b Find the value of k.

A certain amount of money is invested. It earns a fixed rate of interest every year.

The value after t years is p thousand dollars.

c Find the initial value of the investment and the annual rate of interest.

Mathematics in life and work: Group discussion

You are a seismologist interested in the energy released when an earthquake occurs, because this will be connected with the amount of damage the earthquake can cause.

The magnitude of an earthquake is measured on the Richter scale.

This table gives the quantity of energy released for different magnitudes.

Magnitude (M)	3	4	5	6	7
Energy (E joules)	2.0×10^9	6.3×10^{10}	2.0×10^{12}	6.3×10^{13}	2.0×10^{15}

1 Find a formula for the energy released, E, in terms of the magnitude, M.

2 Extend the table to give the energy released for intermediate values such as 3.5 and 4.5.

3 One of the largest earthquakes ever recorded reached approximately 9.5 on the Richter scale. Use your model to estimate how much energy was released.

SUMMARY OF KEY POINTS

> $y = a^x \Leftrightarrow \log_a y = x$. The base of the logarithm is a.

> Three laws of logarithms:

> > $\log_a xy = \log_a x + \log_a y$

> > $\log_a \dfrac{x}{y} = \log_a x - \log_a y$

> > $\log_a x^k = k \log_a x$

> $y = e^x \Leftrightarrow \ln y = x$, where ln means a logarithm to base e.

> The gradient at any point on the curve $y = e^x$ is the y-coordinate at that point.

> Relationships such as $y = ax^n$ or $y = kb^x$ can be transformed to linear form by using logarithms.

EXAM-STYLE QUESTIONS

1 Here is a graph of $y = 1.5^x$.

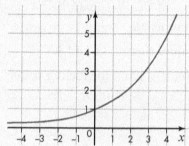

  a Use the graph to find an approximate solution to the equation $1.5^x = 4$.

  b Sketch the graph of $y = 1.5^{-x}$.

  c Write down the coordinates of the point of intersection of $y = 1.5^x$ and $y = 1.5^{-x}$.

2 Solve the equation $\log_5 (3t + 8) = 2.1$, giving the answer correct to 3 significant figures.

3 The equation of a curve is $y = e^{0.5x + 3}$. The curves passes through the points $(-2, k)$ and $(h, 100)$.

  a Find the coordinates of any points where the curve intersects the axes.

  b Find the value of k.

  c Find the value of h.

C 4 a and b are positive integers.

  a Find the value of $\log_{a^2} a$.

  b Prove that $\log_a b \equiv \dfrac{1}{\log_b a}$.

(C) 5 Use logarithms to show that the solution of the inequality $2^x < 3^{2x-1}$ is $x > \dfrac{\ln 3}{\ln 4.5}$.

6 Use logarithms to solve the equation $4^{2x-1} = 5^{3x+1}$, giving your answer correct to 3 significant figures.

(PS) 7 **a** Solve the equation $\ln(35 + 8x) - 2\ln x = \ln 3$.

    b Hence solve the equation $\ln(35 + 8e^{2x+1}) - 2\ln(e^{2x+1}) = \ln 3$, giving your answer correct to 3 significant figures.

8 The variables x and y satisfy the equation $y = Ae^{kx}$, where A and k are constants. It is known that the graph of $y = Ae^{kx}$ passes through the coordinates $(0, 375)$ and $(400, 945)$.

    a Show that $A = 375$.

    b Find the value of k.

    c Find the value of x when $y = 540$.

9 The equation of a curve is $y = 20x^n$, where n is a constant. The point $(15, 590)$ is on the curve. Calculate the value of n, giving your answer correct to 3 significant figures.

(PS) 10 It is given that $\ln(15 - 2x) = \ln(31x - 12) - 2\ln x$.

    a Show that $2x^3 - 15x^2 + 31x - 12 = 0$.

    b By first using the factor theorem, factorise $2x^3 - 15x^2 + 31x - 12$ completely.

    c Hence solve the equation $\ln(15 - 2x) = \ln(31x - 12) - 2\ln x$.

(C) 11 Solve the equation $6|2^x - 3| = 3 \times 2^{x+1}$, giving your answer in the form $x = \dfrac{\ln A}{\ln B} - 1$, where A and B are positive integers.

12

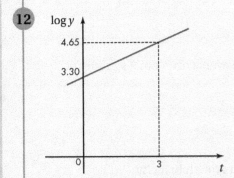

The variables t and y satisfy the equation $y = ka^t$, where k and a are constants. The graph of $\log y$ against t is a straight line passing through the points $(0, 3.30)$ and $(3, 4.65)$, as shown in the diagram. Find the values of k and a, correct to 3 significant figures.

13 **a** Solve the equation $e^x = 4e^{-x}$.

    b Solve the equation $e^x - 3 = 4e^{-x}$.

14 Solve the equation $2^{2x} + 2^3 = 2^{x+3}$.

15 Solve the equation $3^x + \dfrac{4}{3^x} = 5$.

16 Solve the equation $3^x + 3^{x+2} = 3^{x+1} + 20$.

PS **17** The first two terms of a geometric progression are 40 and 39.

MM

   a Find the third term.

   b Find the sum to infinity, S_∞.

   c Given that the sum of the first n terms is S_n and that $S_\infty - S_n < 1$, show that $n \geqslant 292$.

MM **18** The line $y = \dfrac{1}{4}e^{2x}$ meets the line $y = 4$ at the point P.

   a Show that the coordinates of P are $(\ln 4, 4)$.

   b Find the equation of the tangent to $y = \dfrac{1}{4}e^{2x}$ at P.

   c Show that the area of the triangle formed by the tangent and the coordinate axes is
$$\left(\ln\frac{16}{e}\right)^2.$$

19 Two positive variables, s and t, satisfy the equation $s^2 = at^n$, where a and n are constants. The table shows values of $\ln t$ and $\ln s$.

$\ln t$	2.303	4.094
$\ln s$	2.364	2.722

   a Find the value of n.

   b Find the value of a.

   c Find the value of t when $s = 13$.

PS **20** It is given that $2\log_2 (x + 4) - \log_2 x = 5$.

   a Show that $x^2 - 24x + 16 = 0$.

   b Solve the equation $2\log_2 (x + 4) - \log_2 x = 5$, giving your answers correct to 3 significant figures.

Mathematics in life and work

You are a planning consultant and you are looking at population growth in a city.

The current annual rate of growth of the population is 1.5%.

1 If the current rate of growth is unchanged, how many years will it be until the population has increased by 20%?

2 What constant rate of growth would be needed for the population to be less than 10% greater in 20 years' time?

3 TRIGONOMETRY

Mathematics in life and work

Waves are all around us – for example, in water, electricity, light and sound.

One of the uses of trigonometry is as a mathematical tool for analysing waves to investigate their structure. It may be useful to analyse waves in a variety of careers – for example:

> If you were an electrical engineer, you could use trigonometry to combine voltages or currents from different sources.

> If you were a civil engineer, you might use trigonometry to analyse the forces in structures such as bridges and cranes. You would need to make sure that the forces are not large enough to damage the structure.

> If you were a medical technician, you could use trigonometry to analyse the results of magnetic resonance imaging when looking for evidence of disease in a human body.

LEARNING OBJECTIVES

You will learn how to:

> use the expansions of $\sin (A \pm B)$, $\cos (A \pm B)$ and $\tan (A \pm B)$

> use the formulae for $\sin 2A$, $\cos 2A$ and $\tan 2A$

> use the expression of $a \sin \theta + b \cos \theta$ in the forms $R \sin (\theta \pm \alpha)$ and $R \cos (\theta \pm \alpha)$

> understand the relationship of the secant, cosecant and cotangent functions to cosine, sine and tangent

> use the properties and graphs of all six trigonometric functions for angles of any magnitude

> use trigonometric identities for the simplification and exact evaluation of expressions, in particular, $\sec^2 \theta \equiv 1 + \tan^2 \theta$ and $\csc^2 \theta \equiv 1 + \cot^2 \theta$.

LANGUAGE OF MATHEMATICS

Key words and phrases you will meet in this chapter:

> cosecant, cotangent, double angle formulae, secant

PREREQUISITE KNOWLEDGE

You should already know how to:

> find the exact values of trigonometric ratios of any multiple of 30° or 45°

> use the identities $\tan\theta = \dfrac{\sin\theta}{\cos\theta}$ and $\sin^2\theta + \cos^2\theta = 1$

> sketch graphs of sine, cosine and tangent functions

> solve simple trigonometric equations.

You should be able to complete the following questions correctly:

1   Find the exact value of $\sin 600°$.

2   Sketch the graph of $y = 3\sin 2x$ for $0 \leqslant x \leqslant 360°$.

3   Solve the equation $2\cos(x + 0.3) = 0.8$ for $-\pi \leqslant x \leqslant \pi$ radians.

4   Show that $\sin x \tan x \equiv \dfrac{(1 + \cos x)(1 - \cos x)}{\cos x}$.

3.1 Addition and subtraction formulae

Is there a connection between $\sin A$, $\sin B$ and $\sin(A + B)$?

> **Stop and think**   Try values for A and B to check that $\sin(A + B) \neq \sin A + \sin B$.

Look at this diagram.

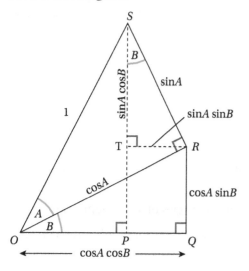

The length of OS is 1 and angle SOP is $A + B$.

From triangle OSP, you can see that

$$PS = OS\sin(A + B) = \sin(A + B) \qquad\qquad ①$$

and

$$OP = OS\cos(A + B) = \cos(A + B). \qquad\qquad ②$$

Now find other expressions for PS and OP.

From triangle *ORS*.

$SR = OS \sin A = \sin A$

and

$OR = OS \cos A = \cos A$.

These are marked on the diagram.

From triangle *OQR*,

$RQ = OR \sin B = \cos A \sin B$ ③

and

$OQ = OR \cos B = \cos A \cos B$.   ④

Now look at triangle *STR*.

Angle $RST = B$

$ST = SR \cos B = \sin A \cos B$ ⑤

and

$RT = SR \sin B = \sin A \sin B$.   ⑥

Finally, you can see from the diagram that $PT = RQ$ and $TR = PQ$.

So $PS = ST + PT = ST + RQ$.

Combine ①, ③ and ⑤.

$\sin (A + B) \equiv \sin A \cos B + \cos A \sin B$ ⑦

Also $OP = OQ - PQ = OQ - TR$.

Combine ②, ④ and ⑥.

$\cos (A + B) \equiv \cos A \cos B - \sin A \sin B$ ⑧

⑦ and ⑧ are important identities. They show how the sine and cosine of the sum of two angles are related to the sines and cosines of the separate angles.

They are valid whether the angles are measured in degrees or in radians.

Example 1

Find an **exact** expression for $\sin 75°$.

Solution

$\sin 75° = \sin (45° + 30°)$

Use the addition formula:

$\sin (45° + 30°) = \sin 45° \cos 30° + \cos 45° \sin 30°$

$$= \frac{1}{\sqrt{2}} \times \frac{\sqrt{3}}{2} + \frac{1}{\sqrt{2}} \times \frac{1}{2}$$

$$= \frac{\sqrt{3} + 1}{2\sqrt{2}}$$

You can derive other formulae from these two formulae.

$\sin(A + B) \equiv \sin A \cos B + \cos A \sin B$

If you replace B with $-B$ you get

$\sin(A - B) \equiv \sin A \cos(-B) + \cos A \sin(-B)$.

But $\cos(-B) \equiv \cos B$ and $\sin(-B) \equiv -\sin B$.

> Remember the symmetries of the sine and cosine graphs.

Hence

$\sin(A - B) \equiv \sin A \cos B - \cos A \sin B$.

In a similar way, you can show that

$\cos(A - B) \equiv \cos A \cos B + \sin A \sin B$.

Stop and think Derive the formula for $\cos(A - B)$ by replacing B with $-B$ in the formula for $\cos(A + B)$.

You can use these formulae to find formulae for $\tan(A + B)$ and $\tan(A - B)$.

$\tan(A + B) = \dfrac{\sin(A + B)}{\cos(A + B)}$

$= \dfrac{\sin A \cos B + \cos A \sin B}{\cos A \cos B - \sin A \sin B}$

Divide every term in the numerator and denominator by $\cos A \cos B$ and use the facts that

$\dfrac{\sin A}{\cos A} = \tan A$ and $\dfrac{\sin B}{\cos B} = \tan B$.

The formula simplifies to

$\tan(A + B) \equiv \dfrac{\tan A + \tan B}{1 - \tan A \tan B}$.

In a similar way, you can prove that

$\tan(A - B) \equiv \dfrac{\tan A - \tan B}{1 + \tan A \tan B}$.

Stop and think Show how to deduce the formula for $\tan(A - B)$ from the appropriate sine and cosine formulae.

KEY INFORMATION

> $\sin(A \pm B) \equiv \sin A \cos B \pm \cos A \sin B$

> $\cos(A \pm B) \equiv \cos A \cos B \mp \sin A \sin B$

> $\tan(A \pm B) \equiv \dfrac{\tan A \pm \tan B}{1 \mp \tan A \tan B}$

Be careful to use the correct signs. It is useful to remember these formulae.

Example 2

Simplify $\cos\left(A + \dfrac{\pi}{4}\right) - \cos\left(A - \dfrac{\pi}{4}\right)$.

Solution

Use the formulae for $\cos(A + B)$ and $\cos(A - B)$.

$\cos\left(A + \dfrac{\pi}{4}\right) - \cos\left(A - \dfrac{\pi}{4}\right)$

$= \cos A \cos\dfrac{\pi}{4} - \sin A \sin\dfrac{\pi}{4} - \left(\cos A \cos\dfrac{\pi}{4} + \sin A \sin\dfrac{\pi}{4}\right)$

$$= -2\sin A \sin\frac{\pi}{4}$$

$$= -2 \times \frac{1}{\sqrt{2}}\sin A$$

$$= -\sqrt{2}\sin A$$

Exercise 3.1A

1 Show that $\cos 75° = \dfrac{\sqrt{3}-1}{2\sqrt{2}}$.

2 Find the exact value of $\tan\dfrac{\pi}{12}$.

3 Simplify $\sin 2A\cos 3A + \sin 3A\cos 2A$.

4 Find the exact value of $\cos 10° \cos 20° - \sin 10° \sin 20°$.

5 Show that $\sin A + \cos A \equiv \sqrt{2}\sin\left(A + \dfrac{\pi}{4}\right)$.

6 **a** Simplify $\sin\left(x - \dfrac{\pi}{6}\right) + 2\sin\left(x + \dfrac{\pi}{6}\right)$.

   b Hence find the exact value of $\sin\dfrac{\pi}{12} + 2\sin\dfrac{5\pi}{12}$.

7 **a** Show that $\tan A + \tan B \equiv \dfrac{\sin(A+B)}{\cos A\cos B}$.

   b Find a similar expression for $\tan A - \tan B$.

(PS) **8** **a** Show that $\sin(A+B) + \sin(A-B) \equiv 2\sin A\cos B$.

   b Hence show that $\sin x + \sin y \equiv 2\sin\left(\dfrac{x+y}{2}\right)\cos\left(\dfrac{x-y}{2}\right)$.

9 $A + B = \dfrac{\pi}{4}$.

   a Show that $\tan A = \dfrac{1-\tan B}{1+\tan B}$.

   b Find a similar formula for $\tan B$.

10 **a** Simplify $\cos(A+B) - \cos(A-B)$.

   b Show that $\cos\alpha - \cos\beta \equiv -2\sin\left(\dfrac{\alpha+\beta}{2}\right)\sin\left(\dfrac{\alpha-\beta}{2}\right)$.

11 a Prove that $\dfrac{\sin(A - B)}{\cos A \cos B} \equiv \tan A - \tan B$.

b Prove that $\dfrac{\sin(A - B)}{\cos A \cos B} + \dfrac{\sin(B - C)}{\cos B \cos C} + \dfrac{\sin(C - A)}{\cos C \cos A} \equiv 0$.

12 a $\sin\left(x + \dfrac{\pi}{6}\right) \equiv a \sin x + b \cos x$.

Find the values of a and b.

b $\sin x + \sin\left(x + \dfrac{\pi}{6}\right) = \sin\left(x + \dfrac{\pi}{3}\right)$.

Show that $\tan x = 2 - \sqrt{3}$.

13 Solve these equations.

a $\sin(\theta - 40°) = \cos(\theta - 20°)$ $0° \leqslant \theta \leqslant 360°$

b $\tan x + \tan(x + 45°) = 2$ $0° \leqslant \theta \leqslant 360°$

3.2 Double angle formulae

If you replace B with A in the addition formulae you get the following **double angle formulae**:

$\sin 2A \equiv 2 \sin A \cos A$

$\cos 2A \equiv \cos^2 A - \sin^2 A$

$\tan 2A \equiv \dfrac{2 \tan A}{1 - \tan^2 A}$

> **Stop and think**   Show how to derive each of these from the addition formulae.

The formula for $\cos 2A$ can be written in different ways.

Remember the formula derived from Pythagoras' theorem:

$\sin^2 A + \cos^2 A = 1$.

If you rearrange this as $\sin^2 A = 1 - \cos^2 A$ and substitute into the formula for $\cos 2A$ you get

$\cos 2A = \cos^2 A - (1 - \cos^2 A)$

or

$\cos 2A = 2\cos^2 A - 1$.

Alternatively, substitute $\cos^2 A = 1 - \sin^2 A$ and get

$\cos 2A = (1 - \sin^2 A) - \sin^2 A$

or $\cos 2A = 1 - 2 \sin^2 A$.

> **KEY INFORMATION**
>
> ❯ $\sin 2A \equiv 2 \sin A \cos A$
>
> ❯ $\cos 2A \equiv \cos^2 A - \sin^2 A$ or $2\cos^2 A - 1$ or $1 - 2\sin^2 A$
>
> ❯ $\tan 2A \equiv \dfrac{2 \tan A}{1 - \tan^2 A}$
>
> It is useful to remember these identities.

Example 3

Solve the equation $\cos 2x = \sin x$ for $0 < x \leqslant 2\pi$; x in radians.

Solution

Use the identity $\cos 2x = 1 - 2\sin^2 x$.

Then $1 - 2\sin^2 x = \sin x$.

Rearrange: $2\sin^2 x + \sin x - 1 = 0$.

This is a quadratic equation in $\sin x$.

Factorise: $(2\sin x - 1)(\sin x + 1) = 0$.

Hence $\sin x = \frac{1}{2}$ or -1.

If $\sin x = \frac{1}{2}$, then $x = \frac{\pi}{6}$ or $\frac{5\pi}{6}$.

If $\sin x = -1$, then $x = \frac{3\pi}{2}$.

There are three solutions in the interval given.

Example 4

Show that $\cos 3A \equiv 4\cos^3 A - 3\cos A$.

Solution

$\cos 3A \equiv \cos(2A + A)$

$\qquad \equiv \cos 2A \cos A - \sin 2A \sin A$

$\qquad \equiv (2\cos^2 A - 1)\cos A - 2\sin A \cos A \sin A$ ⟵ Using the double angle formulae.

$\qquad \equiv 2\cos^3 A - \cos A - 2\cos A \sin^2 A$

$\qquad \equiv 2\cos^3 A - \cos A - 2\cos A(1 - \cos^2 A)$ ⟵ Using $\sin^2 A + \cos^2 A \equiv 1$.

$\qquad \equiv 2\cos^3 A - \cos A - 2\cos A + 2\cos^3 A$

$\qquad \equiv 4\cos^3 A - 3\cos A$

Exercise 3.2A

1 Show that $\frac{1}{2}\sin x \equiv \sin\frac{x}{2}\cos\frac{x}{2}$.

2 **a** Show that $\cos^2 \theta \equiv \frac{1}{2}(1 + \cos 2\theta)$.

    b Hence find an exact expression for $\cos 15°$.

PS 3 **a** Find an expression for $\sin^2 \theta$ in terms of $\cos 2\theta$.

    b Hence show that $\sin\frac{\pi}{8} = \frac{\sqrt{2 - \sqrt{2}}}{2}$.

4 Solve the equation $\sin 2\theta = 1.5 \sin \theta$ for $0° < \theta < 360°$.

5 Show that $\cos x = \frac{1}{2} (\sin x \sin 2x + 2 \cos^3 x)$.

6 Show that $\cos 2\theta + 2 \cos \theta + 1 \equiv 2\cos \theta (\cos \theta + 1)$.

7 **a** If $\tan \theta = \frac{1}{4}$, find the exact value of $\tan 2\theta$.

    b If $\tan 2\theta = 2$ and θ is acute, find the exact value of $\tan \theta$.

8 Solve the equation $\tan 2\theta = 4\tan \theta$ for $0 < \theta < \frac{\pi}{2}$.

    Give your answer to 3 d.p.

9 **a** Sketch a graph to show that the equation $3 \cos 2\theta = \cos \theta - 2$ has two solutions in the domain $0° < \theta < 180°$.

    b Solve the equation $3 \cos 2\theta = \cos \theta - 2$ for $0° < \theta < 180°$.

10 **a** If $x = \frac{\pi}{8}$ show that $2 \tan x = 1 - \tan^2 x$.

    b Solve the quadratic equation in **part a** to find the exact value of $\tan \frac{\pi}{8}$.

PS **11** Show that $\sin 3A \equiv 3 \sin A - 4 \sin^3 A$.

12 Solve the following equations.

    a $2 \sin 2x = \cos x$ **b** $2 \cos 2x = \sin x + 1$ for $0° \leqslant x \leqslant 180°$

13 **a** Show that $\cos 4A \equiv 8 \cos^4 A - 8 \cos^2 A + 1$.

    b By making the substitution $x = \cos A$, find the four solutions of the equation $8x^4 - 8x^2 + 1 = 0$ for $0 < A < \pi$.

14 **a** Show that $\tan 3A = \dfrac{3\tan A - \tan^3 A}{1 - 3\tan^2 A}$.

    b Solve the equation $1 + \tan^3 A = 3 \tan A + 3 \tan^2 A$ for $0° \leqslant A \leqslant 180°$.

15 **a** Show that $\tan(A + 45°) \equiv \dfrac{1 + \tan A}{1 - \tan A}$.

    b Solve the equation $\tan 2x = \tan (x + 45°) - 2$ for $0 < x < 180°$.

16 Write $t = \tan \frac{A}{2}$.

    a Show that $\tan A = \dfrac{2t}{1 - t^2}$.

    b Show that $\dfrac{2t}{1 + t^2} = \sin A$.

    c Write $\cos A$ in terms of t.

Mathematics in life and work: Group discussion

In a hospital, electronic equipment is used to monitor a patient's breathing. A typical breathing rate is 15 breaths per minute.

The rate at which breath is entering or leaving the body, in ml s^{-1}, can be modelled by a sine wave with a period of 4 s.

1 Sketch a graph showing the rate at which breath is entering or leaving the body ($y \, \text{ml s}^{-1}$) against time (t s).

2 Assume the amplitude is a constant $a \, \text{ml s}^{-1}$. Find the equation for y in terms of t.

3 A typical intake of air in one breath when the patient is at rest is 500 ml. How can you use this fact to find the value of the constant a?

4 A breathing rate above 20 breaths per minute is hyperventilation and can be dangerous. How will the graph change if the rate is 20 breaths per minute but the total intake of air in each breath is unchanged?

3.3 The expression $a \sin \theta + b \cos \theta$

An expression of the form $a \cos \theta + b \sin \theta$ can be written in the form of a single sine wave.

For example, here is a graph of $y = 5\sin \theta + 2\cos \theta$.

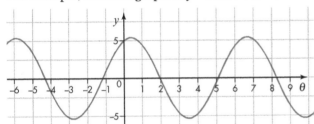

It looks like a sine wave.

Assume the equation can be written in the form $y = r \sin (\theta + \alpha)$, where r and θ are numbers to be found.

That means $5\cos \theta + 2\sin \theta \equiv r \sin (\theta + \alpha)$.

Use the addition formula.

$5\cos \theta + 2\sin \theta \equiv r (\sin \theta \cos \alpha + \cos \theta \sin \alpha)$

$5\cos \theta + 2\sin \theta \equiv (r \sin \alpha) \cos \theta + (r \cos \alpha) \sin \theta$

If these are to be identical, then the coefficients of $\sin \theta$ and $\cos \theta$ must be the same.

So you want to find r and α so that

$r \sin \alpha = 5$ and $r \cos \alpha = 2$.

Divide one expression by the other.

$$\frac{r \sin \alpha}{r \cos \alpha} = \frac{5}{2}$$

> The r cancels and
> $\dfrac{\sin \alpha}{\cos \alpha} \equiv \tan \alpha$.

$\tan \alpha = 2.5$ so $\alpha = 1.19$

To find r, square both equations and add.

$r^2 \sin^2 \alpha + r^2 \cos^2 \alpha = 25 + 4$

$r^2 = 29$ so $r = \sqrt{29} = 5.385$

Put these values in the original identity.

$5 \cos \theta + 2 \sin \theta \equiv 5.385 \sin (\theta + 1.19)$

If you look at the graph above you can check that the amplitude is

5.385 and the graph is a translation of $y = 5.385 \sin \theta$ by $\begin{pmatrix} -1.19 \\ 0 \end{pmatrix}$.

You can generalise this method.

$a \cos \theta + b \sin \theta \equiv r \sin (\theta + \alpha)$.

Expand $\sin (\theta + \alpha)$ and equate coefficients to get

$\alpha = \tan^{-1}\left(\dfrac{a}{b}\right)$ and $r = \sqrt{a^2 + b^2}$.

> Although r could be found by substituting α back into one of the original equations, the 'square and add' method is preferred because it reduces the chance of a follow-through error if α is incorrect.

> Using $\sin^2 \alpha + \cos^2 \alpha \equiv 1$.

Example 5

Solve the equation $5 \cos \theta + 2 \sin \theta = 4$ for $0 < \theta < 2\pi$; θ in radians.

Solution

First write $5 \cos \theta + 2 \sin \theta$ in the form $r \sin (\theta + \alpha)$.

You have already done that above by expanding the expression and equating coefficients.

The equation becomes

$5.385 \sin (\theta + 1.19) = 4$

$\quad \sin (\theta + 1.19) = 0.7428$

$\sin^{-1} 0.7428 = 0.84$ so $\theta + 1.19 = 0.84$ or $\pi - 0.84$ or $2\pi + 0.84$ or $\ldots$

So $\theta = -0.35$ or 1.11 or 5.93.

The first value is out of range so the two answers are $\theta = 1.11$ or 5.93.

> You can use the graph to check that these look correct.

$r \sin (\theta + \alpha)$ is not the only way in which $a \cos \theta + b \sin \theta$ can be rewritten.

$r \sin (\theta - \alpha)$, $r \cos (\theta + \alpha)$ and $r \cos (\theta - \alpha)$ could all be used instead.

Whichever form is used, follow the same procedure – expand the sum or difference and then equate coefficients.

Exercise 3.3A

1   **a**   Write $\cos \theta + \sqrt{3} \sin \theta$ in the form $r \sin (\theta + \alpha)$.

     b   Sketch the graph of $y = \cos \theta + \sqrt{3} \sin \theta$ for $0 \leqslant \theta \leqslant 2\pi$ radians.

2   **a**   Write $3 \sin \theta + 4 \cos \theta$ in the form $r \sin (\theta + \alpha)$, where α is in degrees.

     b   Solve the equation $3 \sin \theta + 4 \cos \theta = 5$ for $0° \leqslant \theta \leqslant 360°$.

3 **a** Write $\cos \theta + \sin \theta$ in the form $r \sin (\theta + \alpha)$, where α is in radians.

b Write $\cos \theta - 7 \sin \theta$ in the form $r \cos (\theta + \alpha)$, where α is in degrees.

c Solve the equation $\cos \theta - 7 \sin \theta = -5$ for $0° \leqslant \theta \leqslant 90°$.

4 **a** Write $8 \sin x - 6 \cos x$ in the form $r \sin (x - \alpha)$.

b Work out the smallest possible value of $8 \sin x - 6 \cos x$.

5 **a** Write $12 \sin \theta + 5 \cos \theta$ in the form $r \sin (\theta + \alpha)$.

b Find the smallest positive solution, in radians, of the equation $12 \sin \theta + 5 \cos \theta = 8$.

6 **a** Write $10 \cos \theta - 12 \sin \theta$ in the form $r \cos (\theta + \alpha)$.

b Solve the equation $10 \cos \theta - 12 \sin \theta = 5$ for $0 \leqslant \theta \leqslant 2\pi$ radians.

(MM) 7 A moving sculpture consists of two metal rods, OR and RP, of length 6 m and 4 m, respectively, in a vertical plane. Angle ORP is a right angle.

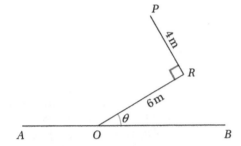

The rods can rotate about O. AOB is a straight line on the ground beneath the sculpture and angle $ROB = \theta$ radians.

a Show that the height of P above the ground, in metres, is $6 \sin \theta + 4 \cos \theta$.

b Find the maximum height of P above the ground.

c Find the two possible values of θ for which P is 7 m above the ground.

(PS) 8 **a** Solve the equation $0.5 \sin \theta + 0.4 \cos \theta = 0.2$ for $0 \leqslant \theta \leqslant 2\pi$ radians.

b Prove that the equation $0.5 \sin \theta + 0.4 \cos \theta = 0.7$ has no solution.

9 **a** Show that $\sin x + \sin (x + 10°) = 1.985 \sin x + 0.174 \cos x$.

b Solve the equation $\sin x + \sin (x + 10°) = 1$ for $0 \leqslant x° \leqslant 180°$.

10 $\cos \left(\theta + \dfrac{\pi}{4} \right) \equiv a \cos \theta + b \sin \theta$

a Find the exact values of a and b.

b Solve the equation $\cos \theta + 2 \cos \left(\theta + \dfrac{\pi}{4} \right) = 2$.

Mathematics in life and work: Group discussion

You are an electrical engineer looking at the combined result of different alternating currents, represented by sine waves or cosine waves. Currents often have the same period but they are out of phase. This means that there is a constant difference between the peaks of the waves representing different currents. This is called the phase difference.

1   Two currents are represented by the curves $y = a \sin x$ and $y = a \cos x$, where x is in radians. How can you show that the phase difference is $\frac{\pi}{2}$?

2   Consider the combined current $y = a \sin x + a \cos x$. What is the phase difference between this and the two initial currents?

3   Consider the combined current $y = a \sin x + b \cos x$, where the amplitudes a and b are different. How do the relative sizes of a and b affect the phase differences between the combined current and the initial currents?

3.4 The secant, cosecant and tangent functions

There are three more trigonometric ratios that can be useful.
They are defined in terms of sine, cosine and tangent.

The **secant** is defined as $\sec \theta \equiv \dfrac{1}{\cos \theta}$.

The **cosecant** is defined as $\operatorname{cosec} \theta \equiv \dfrac{1}{\sin \theta}$.

> cosecant can be abbreviated as cosec or csc. In this book cosec is used, but calculators and computer software often use the abbreviation csc.

The **cotangent** is defined as $\cot \theta \equiv \dfrac{1}{\tan \theta}$.

Here are graphs of the three functions, together with the graphs of the functions from which they are derived.

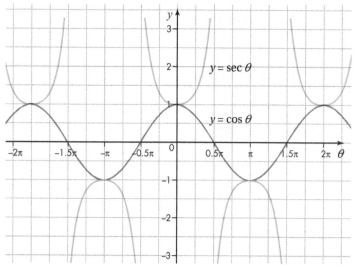

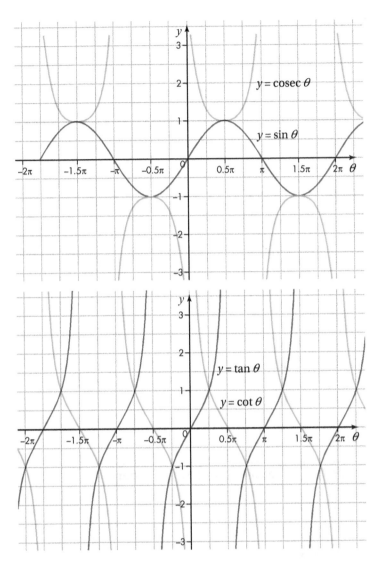

The new functions are not defined if $\sin \theta$, $\cos \theta$ or $\tan \theta = 0$ and values of θ where this is true are excluded from the domain. You can see that the graphs are discontinuous at these values.

Example 6

a   Find the exact value of $\sec \dfrac{\pi}{6}$.

b   Describe the domain of the function $f(x) = 1 + \sec x$.

c   Describe the range of the function $f(x) = 1 + \sec x$.

Solution

a $\sec\dfrac{\pi}{6} = \dfrac{1}{\cos\frac{\pi}{6}} = \dfrac{1}{\frac{\sqrt{3}}{2}} = \dfrac{2}{\sqrt{3}}$

b Exclude values of x for which $\cos x = 0$.

That is $x \neq \pm\dfrac{\pi}{2},\ \pm\dfrac{3\pi}{2},\ \pm\dfrac{5\pi}{2}$, etc.

c The range for $\cos x$ is $-1 \leqslant \cos x \leqslant 1$.

Since $\sec x = \dfrac{1}{\cos x}$ it follows that the range for $\sec x$ is $\sec x \geqslant 1$ and $\sec x \leqslant -1$.

The range for $1 + \sec x$ is $1 + \sec x \geqslant 2$ and $1 + \sec x \leqslant 0$.

Exercise 3.4A

1 Find the exact values of the following.

    a $\sec\dfrac{\pi}{4}$ **b** $\cot\dfrac{\pi}{6}$

    c $\operatorname{cosec}\dfrac{\pi}{3}$ **d** $\operatorname{cosec}\dfrac{2\pi}{3}$

2 Copy this table and insert the exact values.

θ	$\sin\theta$	$\cos\theta$	$\tan\theta$	$\sec\theta$	$\operatorname{cosec}\theta$	$\cot\theta$
$\dfrac{\pi}{6}$			$\dfrac{1}{\sqrt{3}}$			
$\dfrac{\pi}{3}$			$\sqrt{3}$			

3 **a** Solve the equation $\operatorname{cosec} x = 4$ for $0 \leqslant x \leqslant 2\pi$.

    b Solve the equation $\sec x = -3$ for $0 \leqslant x \leqslant 2\pi$.

4 Find the exact values of the following.

    a $\sec 60°$ **b** $\operatorname{cosec} 120°$ **c** $\cot 210°$

5 Solve these equations for $0° \leqslant \theta < 360°$.

    a $\cot(x + 30°) = 0.4$ **b** $\sec(x - 25°) = 2$

    c $3\operatorname{cosec} x = 2\sec x$ **d** $\operatorname{cosec}(2x + 20°) = 1.6$

6 **a** Sketch the graph of the function $f(x) = \operatorname{cosec} 2x$ for $0 \leqslant x \leqslant 2\pi$.

    b Describe the range of the function.

7 **a** On the same axes, sketch graphs of $y = 2\sec x$ and $y = \cot x$ for $0° \leqslant x \leqslant 360°$.

    b Find the coordinates of the points where the graphs cross in the interval $0° \leqslant x \leqslant 360°$.

8 Describe the range of each these functions.

    a $f(x) = 0.5 \sec x$ **b** $f(x) = 0.5 + \sec x$ **c** $f(x) = \sec(x + 0.5)$

 9 Show that $\sec^2 x + \operatorname{cosec}^2 x \equiv 4 \operatorname{cosec}^2 2x$.

10 **a** Prove that $\cot A - \operatorname{cosec} 2A \equiv \cot 2A$.

    b Solve the equation $\cot x = \cot 2x + 4$ for $0 \leqslant x \leqslant 2\pi$.

3.5 More trigonometric identities

Here is a familiar identity: $\sin^2 \theta + \cos^2 \theta \equiv 1$.

This formula can be used to derive related identities involving other trigonometric functions.

Divide each term by $\cos^2 \theta$.

$$\frac{\sin^2 \theta}{\cos^2 \theta} + \frac{\cos^2 \theta}{\cos^2 \theta} \equiv \frac{1}{\cos^2 \theta}$$

$$\left(\frac{\sin \theta}{\cos \theta}\right)^2 + 1 \equiv \left(\frac{1}{\cos \theta}\right)^2$$

$$\tan^2 \theta + 1 \equiv \sec^2 \theta \qquad\qquad\qquad ①$$

Or you could divide each term by $\sin^2 \theta$.

$$\frac{\sin^2 \theta}{\sin^2 \theta} + \frac{\cos^2 \theta}{\sin^2 \theta} \equiv \frac{1}{\sin^2 \theta}$$

$$1 + \cot^2 \theta \equiv \operatorname{cosec}^2 \theta \qquad\qquad\qquad ②$$

Identities ① and ② can be useful in rearranging expressions.

> **KEY INFORMATION**
>
> ❭ $\sec^2 \theta \equiv 1 + \tan^2 \theta$
>
> ❭ $\operatorname{cosec}^2 \theta \equiv 1 + \cot^2 \theta$

Example 7

Solve the equation $\tan x + \sec^2 x = 3$ for $-180° \leqslant x \leqslant 180°$.

Solution

Substitute $\tan^2 x + 1$ for $\sec^2 x$.

$\tan x + \tan^2 x + 1 = 3$

$\tan^2 x + \tan x - 2 = 0$

> Now the equation only contains one trigonometric function.

This is a quadratic expression in $\tan x$.

Factorise: $(\tan x + 2)(\tan x - 1) = 0$.

Either $\tan x + 2 = 0$ or $\tan x - 1 = 0$.

So $\tan x = -2$ or 1.

If $\tan x = -2$, $x = -63.4°$ or $-63.4° + 180° = 116.6°$.

If $\tan x = 1$, $x = 45°$ or $45° - 180° = -135°$.

There are four solutions in the domain $-180° \leqslant x \leqslant 180°$.

Exercise 3.5A

1 $\sec\theta = 3$

Find the exact value of $\tan\theta$.

2 Solve the equation $\operatorname{cosec}^2\theta + \cot^2\theta = 9$ for $0° \leqslant \theta \leqslant 180°$.

(PS) **3** $\sec\theta + \tan\theta = k$

    a Find $\sec\theta - \tan\theta$ in terms of k.

    b Hence find $\sec\theta$ in terms of k.

4 Solve the equation $\sec^2\theta + 2\tan\theta = 4$ for $0 \leqslant \theta \leqslant \pi$ radians.

5 Solve the equation $\tan^2 x = 6\sec x - 10$ for $0° \leqslant \theta \leqslant 180°$.

6 Show that:

    a $\sec x + \tan x \equiv \dfrac{1 + \sin x}{\cos x}$ **b** $\sec x - \tan x \equiv \dfrac{\cos x}{1 + \sin x}$.

(PS) **7** Show that $\tan\theta + \cot\theta \equiv \sec\theta\operatorname{cosec}\theta$.

(PS) **8** Show that $\operatorname{cosec} 2x = \dfrac{1}{2}\operatorname{cosec} x \sec x$.

(PS) **9** Show that $\cot 2x = \dfrac{\cot^2 x - 1}{2\cot x}$.

10 Solve the following equations for $0 < x < \pi$.

    a $\tan x + \cot x = 3$

    b $\tan x + 7\cot x = 5\operatorname{cosec} x$

SUMMARY OF KEY POINTS

❱ $\sin(A \pm B) \equiv \sin A \cos B \pm \cos A \sin B$

❱ $\cos(A \pm B) \equiv \cos A \cos B \mp \sin A \sin B$

❱ $\tan(A \pm B) \equiv \dfrac{\tan A \pm \tan B}{1 \mp \tan A \tan B}$

❱ $\sin 2A \equiv 2 \sin A \cos A$

❱ $\cos 2A \equiv \cos^2 A - \sin^2 A$ or $2\cos^2 A - 1$ or $1 - 2\sin^2 A$

❱ $\tan 2A \equiv \dfrac{2\tan A}{1 - \tan^2 A}$

❱ $a\cos\theta + b\sin\theta$ can be written in the form $r\sin(\theta \pm \alpha)$ or $r\cos(\theta \pm \alpha)$

❱ $\sec\theta \equiv \dfrac{1}{\cos\theta}; \ \operatorname{cosec}\theta \equiv \dfrac{1}{\sin\theta}; \ \cot\theta \equiv \dfrac{1}{\tan\theta}$

❱ $\tan^2\theta + 1 \equiv \sec^2\theta$

❱ $1 + \cot^2\theta \equiv \operatorname{cosec}^2\theta$

EXAM-STYLE QUESTIONS

1 **a** Write down the exact values of:

    i $\cos 30°$ **ii** $\cos 135°$.

  b Hence find the exact value of $\cos 105°$.

2 Show that:

  a $\tan 75° = \dfrac{\sqrt{3}+1}{\sqrt{3}-1}$

  b $\cot 75° = 2 - \sqrt{3}$.

PS **3** **a** Show that $\sin\theta \equiv 2\cos\dfrac{\theta}{2}\sin\dfrac{\theta}{2}$.

  b Hence show that $\sin\theta \equiv 4\cos\dfrac{\theta}{2}\cos\dfrac{\theta}{4}\sin\dfrac{\theta}{4}$.

4 Solve the equation $2\sin^2\theta = \sin\theta$ for $0 \leqslant \theta \leqslant \pi$.

5 Prove that $\dfrac{\cos 2\theta}{\cos^2\theta} + \tan^2\theta \equiv 1$.

PS **6** Solve the equation $\cos^4\theta = 2\sin^2\theta\cos^2\theta$ for $0° \leqslant \theta \leqslant 180°$.

7 θ is an acute angle and $\cos\theta = \frac{4}{5}$.

Show that $\cos\left(\theta + \frac{\pi}{4}\right) = \frac{\sqrt{2}}{10}$.

8 **a** Sketch the graphs of $y = \sec x$ and $y = \cot x$ for $-180° \leqslant \theta \leqslant 180°$.

b Solve the equation $\sec x = \cot x$ for $-180° \leqslant \theta \leqslant 180°$.

9 **a** Express $10\sin\theta + 14\cos\theta$ in the form $R\sin(\theta + \alpha)$, where $R > 0$ and $0 < \alpha < \frac{\pi}{2}$. Give the value of α correct to 2 decimal places.

b Hence solve the equation $10\sin\theta + 14\cos\theta = 15$ for $0 \leqslant \theta \leqslant 2\pi$.

PS 10 **a** Prove that $\cos^2 2A \equiv 4\cos^4 A - 4\cos^2 A + 1$.

b Prove that $\cos 4A \equiv 8\cos^4 A - 8\cos^2 A + 1$.

11 Solve the equation $2\sin x + 1 = \operatorname{cosec} x$ for $0 \leqslant x \leqslant 2\pi$.

C 12 **a** Solve the equation $\tan x + \cot x = 5$ for $0 \leqslant x \leqslant \pi$.

b Find the range of values of k for which $\tan x + \cot x = k$ has a solution.

13 Solve the equation $6\sin^2 x + \cos x = 5$ for $0 \leqslant x \leqslant 2\pi$.

14 **a** Express $5\sin\theta + 12\cos\theta$ in the form $R\cos(\theta - \alpha)$, where $R > 0$ and $0 < \alpha < \frac{\pi}{2}$. Give the value of α correct to 2 decimal places.

b Hence solve the equation $5\sin\theta + 12\cos\theta = 10$ for $0 \leqslant \theta \leqslant 2\pi$.

15 Solve the equation $3\sin\theta + 4\cos\theta = 2$ for $0° \leqslant \theta \leqslant 360°$.

16 Solve the equation $3\sin\theta + \cos 2\theta = 2$ for $0° \leqslant \theta \leqslant 360°$.

17 Given that $\sin(A + B) = 2\sin(A - B)$:

a prove that $\tan A = 3\tan B$

b find the solution of the equation $\sin(\theta + 0.5) = 2\sin(\theta - 0.5)$ for $0 \leqslant \theta \leqslant 2\pi$.

18 **a** Prove that $\cos\left(\frac{\pi}{4} + A\right) + \cos\left(\frac{\pi}{4} - A\right) \equiv \sqrt{2}\cos A$.

b Hence show that $\cos\frac{5\pi}{12} + \cos\frac{\pi}{12} = \frac{1}{2}\sqrt{6}$.

c Find the exact value of $\cos\frac{5\pi}{12} - \cos\frac{\pi}{12}$.

Mathematics in life and work

The sum of two alternating electric currents is represented by the curve $y = 5 \sin \theta + k \cos \theta$, where θ is in radians and k is a positive constant.

The curve can be written in the form $r \sin(\theta + \alpha)$.

1   Find an expression for r in terms of k.

2   If $\alpha = 0.3$, find the value of k.

3   If $5 \sin \theta + k \cos \theta = -4$, find the smallest possible positive value of θ.

4 DIFFERENTIATION

Mathematics in life and work

When quantities change over time, differentiation is the ideal mathematical tool to analyse and study those changes. Differentiation has a wide variety of applications across the physical and natural sciences – for example:

› If you were a molecular biologist, you would use differentiation to analyse changes taking place in a body when a drug is used to combat disease.

› If you were an industrial chemist, you would use differentiation to look at the rates at which chemical reactions take place in different circumstances.

› If you were an environmental scientist, you would use differentiation when studying changes in populations of animals or plants under different conditions.

› If you were an astronomer, you would use differentiation to study the motion of planets, stars and other astronomical bodies.

› If you were an engineer, you would use differentiation to analyse oscillations in any vibrating system, such as a car suspension or a bridge in the wind.

LEARNING OBJECTIVES

You will learn how to:

› differentiate e^x, $\ln x$, $\sin x$, $\cos x$, $\tan x$ and $\tan^{-1} x$

› differentiate products and quotients

› differentiate functions defined parametrically or implicitly.

LANGUAGE OF MATHEMATICS

Key words and phrases you will meet in this chapter:

› implicit equation, parametric equation, product rule, quotient, quotient rule

PREREQUISITE KNOWLEDGE

You should already know how to:

› differentiate x^n for any value of n

› find the equation of a tangent or a normal to a curve

> use the chain rule for differentiation

> use exponential and logarithmic functions.

You should be able to complete the following questions correctly:

1 Differentiate each expression with respect to x.

    a $2x(x^2 - 1)$ **b** $\dfrac{x^2 - 1}{2x}$ **c** $\dfrac{1}{\sqrt{x}}$

2 Find the equation of the normal to the curve $y = x^3 - 6x^2$ at the point $(6, 0)$.

3 If $f(x) = \sqrt{1 + x^2}$, find $f'(x)$.

4 Make x the subject of the formula $y = 5e^{2x - 1}$.

4.1 Differentiating e^x

In this section, you will see that if $y = e^x$, then $\dfrac{dy}{dx} = e^x$.

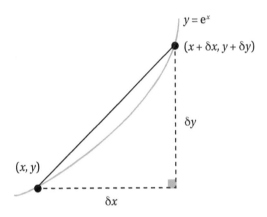

Suppose (x, y) and $(x + \delta x, y + \delta y)$ are two points close together on the graph $y = e^x$.

Then the difference between the two y-coordinates, $\delta y = e^{x + \delta x} - e^x$.

The gradient of the chord joining the two points is

$$\dfrac{\delta y}{\delta x} = \dfrac{e^{x + \delta x} - e^x}{\delta x} = \dfrac{e^x(e^{\delta x} - 1)}{\delta x}.$$

The value of e^x does not change as δx gets smaller so you can take it out of the limit function.

So what is $\lim_{\delta x \to 0} \dfrac{e^{\delta x} - 1}{\delta x}$?

Here is a table of values, rounded to 4 decimal places.

δx	0.1	0.01	0.001
$\dfrac{e^{\delta x} - 1}{\delta x}$	1.0517	1.0050	1.0005

It is reasonable to assume that $\lim_{\delta x \to 0} \frac{e^{\delta x} - 1}{\delta x} = 1$.

Therefore $\frac{dy}{dx} = e^x \lim_{\delta x \to 0} \frac{e^{\delta x} - 1}{\delta x} = e^x$.

This means that for any point on the graph $y = e^x$ the gradient is just the y-coordinate.

If $y = e^x$ then $\frac{dy}{dx} = y$.

You can differentiate $y = e^{kx}$, where k is a constant, using the chain rule.

If $u = kx$ and $y = e^u$, then

$$\frac{dy}{dx} = \frac{dy}{du} \times \frac{du}{dx} = k \times e^u = ke^{kx}.$$

If you multiply the function by a constant, the derivative is multiplied by the constant in the usual way.

If $f(x) = ce^{kx}$ then $f'(x) = cke^{kx}$.

> **KEY INFORMATION**
>
> $y = e^x \Rightarrow \frac{dy}{dx} = e^x$ or
>
> $f(x) = e^x \Rightarrow f'(x) = e^x$

> This result was derived in **Chapter 2 Logarithmic and exponential functions**.

> **KEY INFORMATION**
>
> $y = e^{kx} \Rightarrow \frac{dy}{dx} = ke^{kx}$ or
>
> $f(x) = e^{kx} \Rightarrow f'(x) = ke^{kx}$

Example 1

Find the gradient of the curve $y = 10e^{-0.3x}$ at the point where $x = 5$.

Solution

If $y = 10e^{-0.3x}$, then $\frac{dy}{dx} = -0.3 \times 10e^{-0.3x} = -3e^{-0.3x}$.

When $x = 5$, $\frac{dy}{dx} = -3e^{-0.3 \times 5} = -3e^{-1.5} = -0.669$ to 3 d.p.

Exercise 4.1A

1 Find $\frac{dy}{dx}$ in the following cases.

  a $y = e^{2x}$ **b** $y = e^{-x}$ **c** $y = e^{0.4x}$ **d** $y = e^{4x+2}$

2 Work out $f'(x)$.

  a $f(x) = 4e^{0.5x}$ **b** $f(x) = 100e^{-0.1x}$ **c** $f(x) = 50e^{2x-10}$

3 $f(x) = 4e^{2x}$

  Find:

  a $f'(x)$ **b** $f''(x)$.

(PS) 4   **a** Show that $1.5 = e^{\ln 1.5}$.

  b Hence write 1.5^x in the form e^{kx} where k is a constant.

  c $y = 1.5^x$. Use your answer to **part b** to find $\frac{dy}{dx}$.

    (C) Communication   **(MM)** Mathematical modelling   **(PS)** Problem solving

5

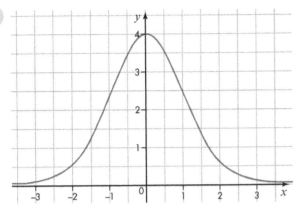

The equation of this curve is $y = 4e^{-0.5x^2}$.

a Find $\dfrac{dy}{dx}$.

b Find the gradient at the point where the x-coordinate is -2.

MM 6 There are bacteria in a colony.

After t hours the number of bacteria is $Ne^{0.1t}$.

a Show that after 3 hours the number of bacteria has increased by 35%.

b Find the rate of growth of the colony after 3 hours.

c After k hours, the rate of growth of the colony is $0.2Nh^{-1}$.

Find the value of k.

7 a $f(x) = e^{2x} + e^{-2x}$

Show that $f''(x) = 4f(x)$.

b Show that the graph of $y = f(x)$ has a minimum point.

PS 8 The equation of a curve is $y = e^x + 4e^{-x}$.

Show that the curve has a minimum point and find its coordinates.

9 The equation of a curve is $y = e^{ax}$ where a is a constant.

a Find the coordinates of the point P on the curve, where the gradient is 1.

b Show the equation of the normal at P can be written as $x + y = k$ and find the value of the constant k.

10 The equation of a curve is $y = e^x - ax$ where $a > 0$.

a Show that the curve has a stationary point and find its coordinates.

b Show that the stationary point is a minimum.

c Find the equation of the tangent where the curve crosses the y-axis.

11 $f(x) = e^x + 3e^{-x}$

a Find the coordinates of the point on the curve $y = f(x)$ where the gradient is 4.

b Show that the minimum value of $f(x)$ is $2\sqrt{3}$.

12 The equation of a curve is $y = ke^x$, where k is a positive constant.
The curve crosses the y-axis at P.

a Find where the tangent at P crosses the x-axis.

b The normal at P meets the x-axis at $(100, 0)$.
Find the value of k.

4.2 Differentiating ln x

Here is a graph of $y = \ln x$.

On the same axes is a graph of $y = e^x$.

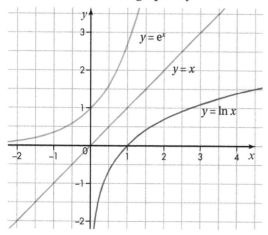

Because finding e to a power and finding ln are inverse functions, the
graph of $y = \ln x$ is a reflection of the graph of $y = e^x$ in the line $y = x$.

You can use this fact to find the gradient of the curve $y = \ln x$ at A.

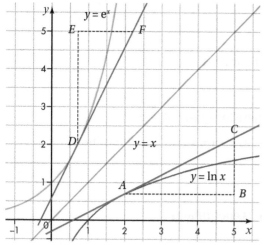

The tangent at A has been drawn and the gradient is $\dfrac{CB}{AB}$.

DEF is a reflection of ABC in the line $y = x$, so $\dfrac{CB}{AB} = \dfrac{FE}{DE}$.

But $\dfrac{FE}{DE} = \dfrac{1}{\dfrac{DE}{FE}} = \dfrac{1}{y\text{-coordinate of }D}$ because $\dfrac{DE}{FE}$ is the gradient of

$y = e^x$ at D, which is just e^x or the y-coordinate.

Because of the reflection, the y-coordinate of D is the x-coordinate of A.

This gives the result that if $y = \ln x$, then $\dfrac{dy}{dx} = \dfrac{1}{x}$.

> **KEY INFORMATION**
>
> $f(x) = \ln x \Rightarrow f'(x) = \dfrac{1}{x}$

Example 2

Differentiate:

a $\ln(2x + 3)$ **b** $\ln x^2$.

Solution

a Use the chain rule.

If $y = \ln(2x + 3)$, then $\dfrac{dy}{dx} = \dfrac{1}{2x+3} \times 2 = \dfrac{2}{2x+3}$.

b Either use the chain rule:

If $y = \ln x^2$, then $\dfrac{dy}{dx} = \dfrac{1}{x^2} \times 2x = \dfrac{2x}{x^2} = \dfrac{2}{x}$.

Or use the properties of logarithms:

If $f(x) = \ln x^2 = 2 \ln x$,

then $f'(x) = 2 \times \dfrac{1}{x} = \dfrac{2}{x}$, which is the same result.

> Using the fact that $\ln a^k = k \ln a$

Example 3

Find the equation of the tangent to the curve $y = \ln(x + e)$ at the point $(0, 1)$.

Solution

Use the chain rule to find $\dfrac{dy}{dx}$.

$\dfrac{dy}{dx} = \dfrac{1}{x+e} \times 1 = \dfrac{1}{x+e}$

At $(0, 1)$, $x = 0$ and $\dfrac{dy}{dx} = \dfrac{1}{e}$.

The equation of the tangent is $y - 1 = \dfrac{1}{e}(x - 0)$.

This can be rearranged as $y = \dfrac{x}{e} + 1$.

> The derivative of $x + e$ is 1.

> An alternative method for the final part of this question is to use $y = mx + c$.
>
> $y = \dfrac{1}{e}x + c$
>
> Substitute $(0, 1) \Rightarrow c = 1$
>
> So $y = \dfrac{1}{e}x + 1$.

Exercise 4.2A

1 Differentiate the following with respect to x.

   a $\ln 3x$ **b** $\ln x^3$ **c** $\ln(x^3 + 2)$

2 Find $\dfrac{dy}{dx}$ in the following.

   a $y = 4\ln x$ **b** $y = \ln 4x$ **c** $y = \ln x^4$

3 **a** On the same axes, sketch the graphs of $y = \ln x$ and $y = \ln 2x$.

   b Find the vector for the translation that maps the graph of $y = \ln x$ onto $y = \ln 2x$.

   c $f(x) = \ln kx$ where k is a positive constant, $x > 0$.

      Show that $f'(x) = \dfrac{1}{x}$.

   d How are your answers to **parts b** and **c** related?

4 This is a graph of the curve $y = \ln(x + 5)$.

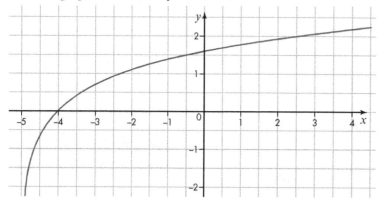

   a Find the equation of the tangent to the curve at the point where it crosses the x-axis.

   b Find the equation of the tangent to the curve at the point where it crosses the y-axis.

PS 5 **a** Show that the point $(e, 1)$ is on the graph of the curve $y = \ln x$.

   b Show that the tangent to the curve $y = \ln x$ at $(e, 1)$ passes through the origin.

   c Show that the normal to the curve $y = \ln x$ at $(e, 1)$ crosses the x-axis at $e + \dfrac{1}{e}$.

PS 6 **a** Find $f'(x)$ when $f(x) = \ln \dfrac{10}{x}$.

   b Find $f'(x)$ when $f(x) = \ln \dfrac{10}{x^2}$.

   c Generalise your results of **a** and **b** to find $f'(x)$ when $f(x) = \ln \dfrac{10}{x^n}$, where n is a positive integer.

PS 7 **a** $y = \log_{10} x$

      Show that $\dfrac{dy}{dx} = \dfrac{1}{x \ln 10}$.

   b Find $\dfrac{dy}{dx}$ when $y = \log_{10} ax^2$.

8 The equation of a curve is $y = x - \ln x$, where $x > 0$.

The curve has a stationary point at S.

a Find the coordinates of S.

b Show that the curve $y = x^n - \ln x$, where n is a positive integer, passes through S.

c Find the gradient of $y = x^n - \ln x$ at S.

d Show that the minimum value of $x^2 - \ln x$ is $\frac{1}{2}(1 + \ln 2)$.

9 P is the point on the curve $y = \ln x$ where the tangent passes through the origin.

a Find the coordinates of P.

Q is the point on $y = \ln x$ where the gradient is e.

b Find the coordinates of Q.

c Find where the tangent at Q crosses the y-axis.

10 The equation of a curve is $y = \ln(ax + b)$, where a and b are positive constants.

The gradient at point P is 1.

a Show that the x-coordinate of P is less than 1.

P is the point $(0, 2)$.

b Find the coordinates of the point where the gradient is $\frac{1}{2}$.

4.3 Differentiating sin x and cos x

Here is a graph of $f(x) = \sin x$ where x is in radians.

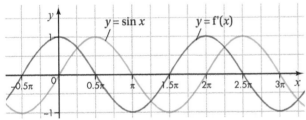

The pink curve is a graph of the gradient $y = f'(x)$.

For example, the gradient at $(0, 0)$ or $f'(0) = 1$.

The gradient at $(0.5\pi, 1)$ or $f'(0.5\pi) = 0$.

The gradient at $(\pi, 0)$ or $f'(\pi) = -1$. •————

The gradient graph looks like the graph of $y = \cos x$. In fact it is identical to it.

> Graph plotting software will often draw the gradient graph for you.

If $f(x) = \sin x$ and x is in radians, then $f'(x) = \cos x$.

Here is a graph of $f(x) = \cos x$ and the gradient $y = f'(x)$.

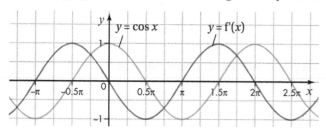

In this case, the gradient graph is a reflection of $y = \sin x$ in the x-axis.

If $f(x) = \cos x$ and x is in radians, then $f'(x) = -\sin x$.

KEY INFORMATION

$f(x) = \sin x \Rightarrow f'(x) = \cos x$

$f(x) = \cos x \Rightarrow f'(x) = -\sin x$

Stop and think   Would this result still be true if the angle was in degrees rather than in radians?

Example 4

Differentiate:

a $2 \sin 4x$ **b** $\sin 3x^2$.

Solution

Use the chain rule in both cases.

a If $f(x) = 2 \sin 4x$, then $f'(x) = 2 \cos 4x \times 4 = 8 \cos 4x$.

b If $f(x) = \sin 3x^2$, then $f'(x) = \cos 3x^2 \times 6x = 6x \cos 3x^2$.

Exercise 4.3A

1 Find $\dfrac{dy}{dx}$ in the following cases.

   a $y = \sin 2x$ **b** $y = \cos (5x - 2)$ **c** $y = \sin (x^2 + 1)$

2 Differentiate the following with respect to x.

   a $10 \sin 0.5x$ **b** $\sin 3x + \cos 6x$ **c** $\cos (x^2 - 3x - 4)$

3 $f(x) = \sin 4x + \cos 4x$

   Show that $f''(x) + 16f(x) = 0$.

4 **a** Show that the derivative of $\sin^2 x$ is $\sin 2x$.

   b Find the derivative of $\cos^2 x$.

   c Explain the relationship between your answers to **parts a** and **b**.

5 This is a graph of $y = e^{\sin x}$.

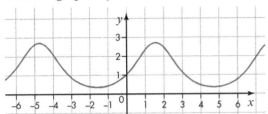

   a Find the gradient at the point $(0, 1)$.

   b Prove that the graph has a stationary point at $\left(\dfrac{\pi}{2}, e\right)$.

(PS) 6 $f(x) = 3 \sin x - \sin^3 x$

   a Show that $f'(x) = 3 \cos^3 x$.

   b Find $f''(x)$.

(PS) 7 a $\sec x = (\cos x)^{-1}$

   Use this fact to show that if $f(x) = \sec x$, then $f'(x) = \sec x \tan x$.

   b If $y = \operatorname{cosec} x$, find an expression for $\dfrac{dy}{dx}$.

(MM) 8 A bob on the end of a long string is a pendulum making small oscillations.

The displacement, y metres, of the bob after t seconds is given by $y = 0.1 \sin 2.4t$.

   a Find the speed of the bob when $t = 1$.

   b Find the position of the bob when the speed is zero.

   c Find the position of the bob when the acceleration is zero.

(PS) 9 $f(x) = \cos \dfrac{1}{x}$, where $x > 0$.

   a Find $f'(x)$.

   b Find the stationary points on the graph of $y = f(x)$.

10 $y = 2 \sin x - 4 \cos x$

   a Find the gradient of the curve when $x = 0$.

   b Find the equation of the tangent to the curve when $x = \dfrac{\pi}{2}$.

   c Show that the curve has a stationary point when x is approximately 2.7.

11 $f(x) = \cos 2x - 2 \cos x$

   a Find $f'(x)$.

   b Find the stationary points when $0 \leqslant x \leqslant \dfrac{\pi}{2}$.

12 $y = x + \cos 2x$

a Find the equation of the normal to the curve where it crosses the y-axis.

b Show that the curve has two stationary points in the interval $0 \leqslant x \leqslant \frac{\pi}{2}$.

c Find the maximum value of y in the interval $0 \leqslant x \leqslant \frac{\pi}{2}$.

13 The equation of a curve is $y = \sin 2\theta + 2 \sin \theta$.

a Find $\frac{dy}{d\theta}$.

b Show that the curve has a maximum point when $\theta = \frac{\pi}{3}$.

14 $y = \sin x + \sin 3x$

Find the equation of the tangent at the point where $x = \frac{\pi}{4}$.

4.4 The product rule

You know how to differentiate the functions $e^{-0.2x}$ and $\sin 4x$.
When you multiply them together you get the product $e^{-0.2x} \sin 4x$.
How can you differentiate a product of two different functions?

Suppose $y = uv$, where u and v are both functions of x.

Suppose δx is a small increase in x that produces changes in y, u, and v of δy, δu and δv, respectively.

This means y changes to $y + \delta y$, u changes to $u + \delta u$ and v changes to $v + \delta v$.

$$\frac{dy}{dx} = \lim_{\delta x \to 0} \frac{\delta y}{\delta x}$$

$$\frac{\delta y}{\delta x} = \frac{(u + \delta u)(v + \delta v) - uv}{\delta x}$$

$$= \frac{uv + u\delta v + v\delta u + \delta u \delta v - uv}{\delta x}$$

$$= \frac{u\delta v + v\delta u + \delta u \delta v}{\delta x}$$

$$= u\frac{\delta v}{\delta x} + v\frac{\delta u}{\delta x} + \delta u \frac{\delta v}{\delta x}$$

$$\frac{dy}{dx} = \lim_{\delta x \to 0} \frac{\delta y}{\delta x} = \lim_{\delta x \to 0} u\frac{\delta v}{\delta x} + \lim_{\delta x \to 0} v\frac{\delta u}{\delta x} + \lim_{\delta x \to 0} \delta u \frac{\delta v}{\delta x}$$

The last term is $\lim_{\delta x \to 0} \delta u \frac{\delta v}{\delta x} = \lim_{\delta x \to 0} \delta u \times \lim_{\delta x \to 0} \frac{\delta v}{\delta x} = 0 \times \frac{dv}{dx} = 0$.

So $\dfrac{dy}{dx} = \lim_{\delta x \to 0} u\dfrac{\delta v}{\delta x} + \lim_{\delta x \to 0} v\dfrac{\delta u}{\delta x}$.

$$\frac{dy}{dx} = u\frac{dv}{dx} + v\frac{du}{dx}$$

This is the **product rule** for differentiation.

> **KEY INFORMATION**
>
> If $f(x) = uv$, where u and v are functions of x, then
> $$f'(x) = u\frac{dv}{dx} + v\frac{du}{dx}.$$

Example 5

This graph shows a damped sine wave. The amplitude is gradually decreasing.

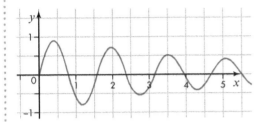

The equation of the curve is $y = e^{-0.2x} \sin 4x$, where $x > 0$.

a Find $\dfrac{dy}{dx}$.

b Find the x-coordinate of the first stationary point.

Solution

a Write $u = e^{-0.2x}$ and $v = \sin 4x$.

Then $\dfrac{du}{dx} = -0.2e^{-0.2x}$ and $\dfrac{dv}{dx} = 4\cos 4x$.

$\dfrac{dy}{dx} = u\dfrac{dv}{dx} + v\dfrac{du}{dx}$

$= e^{-0.2x} \times 4\cos 4x + \sin 4x \times -0.2e^{-0.2x}$

$= e^{-0.2x}(4\cos 4x - 0.2\sin 4x)$

> An alternative method for solving this question would have been to write $y = \dfrac{\sin 4x}{e^{0.2x}}$ and then use the quotient rule, which is covered later in this chapter.

b At a stationary point:

$e^{-0.2x}(4\cos 4x - 0.2\sin 4x) = 0$

$4\cos 4x - 0.2\sin 4x = 0$

> You can cancel $e^{-0.2x}$ because it is always positive and cannot be zero.

Rearrange.

$0.2\sin 4x = 4\cos 4x$

Divide by 0.2.

$\sin 4x = 20\cos 4x$

Divide by $\cos 4x$.

$\tan 4x = 20$

> $\dfrac{\sin 4x}{\cos 4x} \equiv \tan 4x$

Hence $4x = \tan^{-1} 20 = 1.521$.

$x = 0.380$ is the x-coordinate of the first stationary point.

Exercise 4.4A

1 Find $\dfrac{dy}{dx}$.

    a $y = x \sin x$ **b** $y = x \cos x$ **c** $y = x^2 \sin 2x$

2 Differentiate with respect to x.

    a xe^x **b** $(x+1)e^{-x}$ **c** $x^2 e^{2x}$

(PS) 3 $f(x) = \sin x \cos x$

    a Find $f'(x)$ by using the product rule.

    b Find $f'(x)$ by using an identity for $\sin 2x$.

    c Show that your answers to **parts a** and **b** are identical.

4 $y = e^x \sin x$

    a Work out $\dfrac{dy}{dx}$.   **b** Work out $\dfrac{d^2y}{dx^2}$.

5 Differentiate with respect to t.

    a $\sin 10t\, e^{0.2t}$ **b** $\cos t^2\, e^{-1.5t}$ **c** $\sin(2t+1)\, e^{-t^2}$

6 The equation of a curve is $y = x \ln x$, where $x > 0$.

    a Find $\dfrac{dy}{dx}$.

    b Find the gradient at $(1, 0)$.

    c Find the coordinates of the point where the gradient is 2.

    d Find the coordinates of the stationary point.

(PS) 7 **a** The equation of a curve is $y = xe^{-x}$.

    Show that the curve has one stationary point and find its coordinates.

    b The equation of a second curve is $y = x^2 e^{-x}$.

    Show that this curve has two stationary points and find their coordinates.

8 $y = 10x^2\, e^{-x^2}$

    a Find $\dfrac{dy}{dx}$.

    b Find the coordinates of the stationary points.

(PS) 9 The equation of a curve is $y = e^{-0.2x} \sin 2x$, where $x \geqslant 0$.

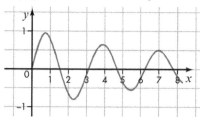

Show that the curve has a stationary point at $x = \dfrac{n\pi}{2} + 0.736$, where n is 0, 1, 2, 3, ...

10 This graph shows the speed of a car (y m s^{-1}) over a 10-second interval.

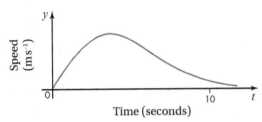

The speed at time t s is given by the formula $y = 15te^{-0.05t^2}$, where $0 \leqslant t \leqslant 10$.

Find the maximum speed of the car.

11 $y = e^x \sin x$

  a Find $\dfrac{dy}{dx}$.

  b Show that $\dfrac{d^2y}{dx^2} = 2\dfrac{dy}{dx} - 2y$.

  c Find the stationary point in the interval $0 \leqslant x \leqslant \pi$ and show that it is a maximum point.

12 The equation of a curve is $y = e^{-x} \sin x$ where $x \geqslant 0$.

  a Show that $\dfrac{d^2y}{dx^2} = -2e^{-x} \cos x$.

  b Show that $\left(\dfrac{\pi}{4}, \dfrac{1}{\sqrt{2}} e^{-\frac{\pi}{4}} \right)$ is a maximum point on the curve.

  c Find the next maximum point on the curve.

  d Show that the values of y at the maximum points form a geometric progression.

13 $y = \sqrt{1 + x} \sin \dfrac{\pi x}{2}$

  a Find $\dfrac{dy}{dx}$.

  b Find the equation of the tangent when $x = 3$.

14 The equation of a curve is $y = x^2\sqrt{20 - x}$ where $0 < x < 20$.
Find the coordinates of the stationary point.

Mathematics in life and work: Group discussion

A sine wave has an equation of the form $y = a \sin \omega t$.

In practice sine waves are often damped. This means that they decay over time and the equation will have the form $y = ae^{-kt} \sin \omega t$, where $k > 0$.

As an engineer you might need to investigate oscillations in a building in the wind or in the parts of a machine.

1   What is the significance of the constants k and ω in the equation above? How would the size of k affect the shape of the curve?

This is the general shape of a damped sine wave.

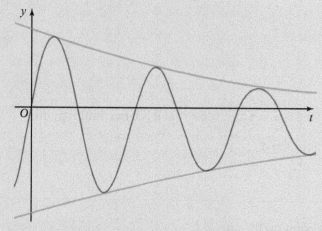

2   What are the equations of the two lines that enclose the curve?

3   Find an expression for $\dfrac{dy}{dx}$. How can you use it to identify the stationary points on the curve? How do the coordinates vary with k and ω?

4   What information could an engineer extract from the graph above if it represented the oscillations of parts of a machine over time?

4.5 The quotient rule

You now know how to differentiate the product of two functions of x.

What if y is the **quotient** of two functions of x, that is, $y = \dfrac{u}{v}$?

Use the same notation as the last section.

$$\delta y = \frac{u + \delta u}{v + \delta v} - \frac{u}{v}$$

Write this as a single fraction.

$$\delta y = \frac{(u + \delta u)v - u(v + \delta v)}{(v + \delta v)v}$$

$$= \frac{uv + v\delta u - uv - u\delta v}{v^2 + v\delta v}$$

$$= \frac{v\delta u - u\delta v}{v^2 + v\delta v}$$

$$\frac{\delta y}{\delta x} = \frac{v\frac{\delta u}{\delta x} - u\frac{\partial v}{\delta x}}{v^2 + v\delta v}$$

$$\frac{dy}{dx} = \lim_{\delta x \to 0}\frac{\delta y}{\delta x} = \frac{\lim_{\delta x \to 0} v\frac{\delta u}{\delta x} - \lim_{\delta x \to 0} u\frac{\delta v}{\delta x}}{\lim_{\delta x \to 0}(v^2 + v\delta v)}$$

$$\frac{dy}{dx} = \frac{v\frac{du}{dx} - u\frac{dv}{dx}}{v^2}$$

This is the **quotient rule** for differentiation.

<table>
<tr><td>

> **KEY INFORMATION**
>
> If $f(x) = \dfrac{u}{v}$, then
>
> $$f'(x) = \frac{v\frac{du}{dx} - u\frac{dv}{dx}}{v^2}.$$

</td></tr>
</table>

> **Example 6**
>
> Find $\dfrac{dy}{dx}$ when $y = \dfrac{1 - x^2}{1 + x^2}$.
>
> **Solution**
>
> $u = 1 - x^2$ and $v = 1 + x^2$.
>
> $\dfrac{du}{dx} = -2x$ and $\dfrac{dv}{dx} = 2x$.
>
> $$\frac{dy}{dx} = \frac{v\frac{du}{dx} - u\frac{dv}{dx}}{v^2}$$
>
> $$= \frac{(1 + x^2) \times (-2x) - (1 - x^2) \times 2x}{(1 + x^2)^2}$$
>
> $$= \frac{-2x - 2x^3 - 2x + 2x^3}{(1 + x^2)^2}$$
>
> $$\frac{dy}{dx} = \frac{-4x}{(1 + x^2)^2}$$

Be very careful with the signs. It is easy to make a mistake.

You know that $\tan x = \dfrac{\sin x}{\cos x}$ so you can use the quotient rule to differentiate $\tan x$.

$\tan x = \dfrac{u}{v}$, where $u = \sin x$ and $v = \cos x$.

$\dfrac{du}{dx} = \cos x$ and $\dfrac{dv}{dx} = -\sin x$

$$\frac{dy}{dx} = \frac{v\frac{du}{dx} - u\frac{dv}{dx}}{v^2}$$

$$= \frac{\cos x \times \cos x - \sin x \times (-\sin x)}{\cos^2 x}$$

$$= \frac{\cos^2 x + \sin^2 x}{\cos^2 x}$$

$\cos^2 x + \sin^2 x = 1$

$$= \frac{1}{\cos^2 x}$$

But you already know that $\sec x = \dfrac{1}{\cos x}$.

Therefore $\dfrac{d}{dx}(\tan x) = \sec^2 x$.

KEY INFORMATION
$f(x) = \tan x \Rightarrow f'(x) = \sec^2 x$

Exercise 4.5A

1 Find $\dfrac{dy}{dx}$.

   a $y = \dfrac{x}{x+1}$
      b $y = \dfrac{x+1}{x^2+1}$
      c $y = \dfrac{x^3}{2x-1}$

2 Find $f'(x)$.

   a $f(x) = \dfrac{\sin x}{x}$
      b $f(x) = \dfrac{x+1}{\cos x}$
      c $f(x) = \dfrac{x^2}{\sin 2x}$

3 Find $\dfrac{dy}{dx}$.

   a $y = \tan ax$
      b $y = \tan(4x+3)$
      c $y = 3\tan(x^2)$

4 Differentiate.

   a $x \tan x$
      b $x^2 \tan 2x$
      c $e^{-x} \tan ax$

(PS) 5 $y = e^{-2x} \sin x$

   a Use the product rule to find $\dfrac{dy}{dx}$.

   b Show that you get the same answer if you write $y = \dfrac{\sin x}{e^{2x}}$ and use the quotient rule.

6 $f(x) = \dfrac{\ln x}{x}$, where $x > 0$

   a Find $f'(x)$.

   b Find $f''(x)$.

   c Find the coordinates of the stationary point of the graph of $y = \dfrac{\ln x}{x}$ when $x > 0$, and show that it is a maximum point.

7 $y = \dfrac{1 - e^{-x}}{1 + e^{-x}}$

   a Find $\dfrac{dy}{dx}$.

   b Find the gradient of the graph of y at the origin.

   c Show that the graph of y has no stationary points.

(PS) 8 $f(x) = \cot x$

   a Show that $f'(x) = -\csc^2 x$.

   b Differentiate $x \cot x$.

9 **a** Show that the derivative of $\dfrac{x}{\sqrt{x^2+1}}$ is $\dfrac{1}{\sqrt[3]{x^2+1}}$.

   b Differentiate $\dfrac{\sqrt{x^2+1}}{x}$.

10 $y = \tan x$

   Show that $\dfrac{dy}{dx} = \dfrac{1}{\cos^2 x}$.

11 $y = \dfrac{x+1}{x-1}$ $x \neq 1$

   a Find $\dfrac{dy}{dx}$.

   b Rewrite the equation to give x as a function of y.

   c Use your answer to **part b** to find $\dfrac{dx}{dy}$.

   d Show that $\dfrac{dx}{dy} = \dfrac{1}{\frac{dy}{dx}}$.

12 $f(x) = \dfrac{x^2}{\cos x}$

   a Find $f'(x)$.

   P is the point on the graph of $y = f(x)$ with an x coordinate of 1.

   b Show that the gradient at P is 6.584 to 3 d.p.

   c Find the equation of the tangent at P.

13 $y = \dfrac{x-1}{\sqrt{x^2+1}}$

   a Differentiate y with respect to x.

   b Find the stationary point on the curve.

   c Find the gradient where the curve crosses each of the axes.

14 The equation of a curve is $y = \dfrac{x}{(x+1)^2}$ $x \neq -1$.

   a Find the coordinates of the stationary point.

   b Show that there is only one point where the gradient is 1 and find its coordinates.

15 $y = \dfrac{1-x^2}{1+x^2}$

   a Show that the curve has a stationary point.

   b Show that $\dfrac{d^2y}{dx^2} = \dfrac{12x^2-4}{(1+x^2)^3}$.

   c Show that the stationary point is a maximum point.

16 $y = \dfrac{\sin x + \cos x}{e^x}$ $x \geqslant 0$

a Show that the gradient is negative for $0 < x < \pi$.

b Show that the stationary points are regularly spaced along the positive x-axis.

c Show that the stationary points alternate between maximum and minimum points.

Mathematics in life and work: Group discussion

Suppose you are an ecologist modelling the growth of a population. This could be a population of animals or plants or other living organisms in a particular environment.

An initial model might be that the population y after time t is given by the formula $y = ae^{kt}$, where $t \geqslant 0$.

1 What is the meaning of the constants a and k in this model?

2 By looking at the rate of growth of the population, explain why this model cannot still be appropriate after a certain amount of time.

A more realistic model is based on the so-called logistic equation, which in its simplest form can be written as $y = \dfrac{e^t}{e^t + 1}$. In this simplified form t can be positive or negative.

3 What is the shape of this curve? Why is it a more appropriate shape?

4 Can you find an expression for $\dfrac{dy}{dt}$ in terms of y? How does the magnitude of $\dfrac{dy}{dt}$ change as t increases from a negative value to a positive value?

4.6 Differentiating $\tan^{-1} x$

The derivative of $\tan^{-1} x$ is rather surprising.

If $y = \tan^{-1} x$, then $x = \tan y$.

$\dfrac{dx}{dy} = \sec^2 y$

Hence

$\dfrac{dy}{dx} = \dfrac{1}{\sec^2 y}$.

Use the fact that $\sec^2 y = 1 + \tan^2 y$.

$\sec^2 y = 1 + x^2$

Therefore $\dfrac{d}{dx}(\tan^{-1} x) = \dfrac{1}{1 + x^2}$.

> **KEY INFORMATION**
>
> $f(x) = \tan^{-1} x \Rightarrow f'(x) = \dfrac{1}{1 + x^2}$

Stop and think

Here is a reminder of the graph of $y = \tan^{-1} x$.

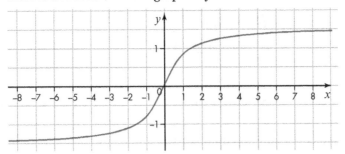

Does the result give values that match the shape of the graph?

What is the maximum gradient?

Example 7

$f(x) = \tan^{-1}(2x + 3)$

a Show that $f(x)$ is an increasing function.

b Find the coordinates of the point on the graph of
$y = \tan^{-1}(2x + 3)$, where the gradient is a maximum.

c Find the x-coordinates of the points on the graph of
$y = \tan^{-1}(2x + 3)$, where the gradient is 0.4.

Solution

a $f(x)$ is an increasing function if $f'(x) \geqslant 0$ for all values of x.

Using the chain rule,

$$f'(x) = \frac{1}{1 + (2x + 3)^2} \times 2 = \frac{2}{1 + (2x + 3)^2}.$$

The denominator is $\geqslant 1$ and positive for all values of x.

Hence $\dfrac{2}{1 + (2x + 3)^2}$ is positive for all values of x and the

function is increasing.

b $f'(x)$ is a maximum when $1 + (2x + 3)^2$ is a minimum.
This happens when $2x + 3 = 0$ and so $x = -1.5$.
Then $y = \tan^{-1} 0 = 0$ and the coordinates are $(-1.5, 0)$.

c If the gradient is 0.4, then $\dfrac{2}{1 + (2x + 3)^2} = 0.4$.

Rearranging:
$1 + (2x + 3)^2 = 5$
$(2x + 3)^2 = 4$
$2x + 3 = \pm 2$
So $x = -0.5$ or -2.5.

PURE MATHEMATICS 3

Exercise 4.6A

1 Differentiate the following with respect to x.

   a $\tan^{-1} ax$ **b** $\tan^{-1}(0.5x - 1)$ **c** $3\tan^{-1}(x^2)$

2 Differentiate:

   a $y = 10\tan^{-1} x$ **b** $y = \tan^{-1} 10x$ **c** $y = \tan^{-1}\frac{x}{10}$.

3 The equation of a curve is $y = \tan^{-1}\frac{x}{a}$ where $a > 0$.

   a Find the coordinates of the point where the gradient is $\frac{1}{2a}$.

   b Find the equation of the normal at this point.

4 $f(x) = x\tan^{-1} x - 0.5\ln(1 + x^2)$

   Find $f'(x)$.

PS 5 $y = \tan^{-1} 2x$

   Find the coordinates of the points where the gradient is 1.

6 Find an expression for $f'(x)$ when:

   a $f(x) = \tan^{-1}\sqrt{x - 1}$ $x > 0$

   b $f(x) = \tan^{-1}\left(\frac{1}{x}\right)$ $x > 0$.

7 Differentiate.

   a $(1 + x^2)\tan^{-1} x$ **b** $\dfrac{\tan^{-1} x}{1 + x^2}$.

8 **a** Find $\displaystyle\int \frac{1}{1 + x^2}\,dx$.

   b Show that if $f(x) = \frac{1}{a}\tan^{-1}\frac{x}{a}$, then $f'(x) = \dfrac{1}{a^2 + x^2}$.

   c Find $\displaystyle\int \frac{1}{25 + x^2}\,dx$.

9 The equation of a graph is $y = \tan^{-1} ax$ where a is a positive constant.

   a Find the gradient at the origin.

   b Find the coordinates of the points where the gradient is $\frac{a}{2}$.

10 Differentiate:

   a $\tan^{-1}(x^3)$ **b** $(\tan^{-1} x)^3$ **c** $x^3\tan^{-1} x$.

4.7 Differentiating parametric equations

Sometimes it is useful to give the equation of a curve by defining x and y in terms of a third variable.

For example, $x = 5 \cos t$, $y = 5 \sin t$ is a **parametric equation** of a curve.

> Graphical software often allows parametric equations to be used.

The variable t is called a **parameter**.

To draw the curve, you need to choose different values for the variable t and work out the values of x and y.

t	0	0.5	1.0	1.5	2.0	2.5	3.0
x	5	4.388	2.702	0.354	−2.081	−4.006	−4.950
y	0	2.397	4.207	4.987	4.546	2.992	0.706

In fact, the curve is a circle.

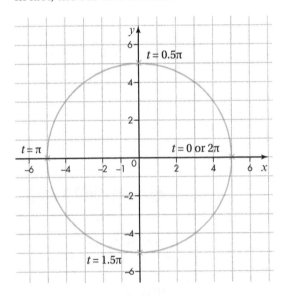

The values of t can be restricted to the interval $0 \leqslant t \leqslant 2\pi$ because after that the same points are generated.

> **Stop and think**
>
> What is the Cartesian equation of this circle?
>
> Why might a parametric equation be more convenient?

Here is another example.

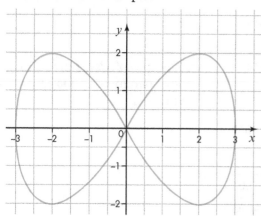

The parametric equation of this curve is $x = 3\cos t$ and $y = 2\sin 2t$ for $0 \leqslant t \leqslant 2\pi$.

For curves that are described parametrically, you can write $\dfrac{\mathrm{d}y}{\mathrm{d}x}$ as a function of the parameter.

If δt is a small change in t, there will be corresponding small changes in δy and δx.

$$\frac{\mathrm{d}y}{\mathrm{d}x} = \lim_{\delta x \to 0} \frac{\delta y}{\delta x}$$

$$\frac{\delta y}{\delta x} = \frac{\delta y}{\delta t} \div \frac{\delta x}{\delta t}$$

$$\lim_{\delta t \to 0} \frac{\delta y}{\delta x} = \lim_{\delta t \to 0} \frac{\delta y}{\delta t} \div \lim_{\delta t \to 0} \frac{\delta x}{\delta t}$$

But as $\delta t \to 0$, then $\delta x \to 0$.

So $\displaystyle\lim_{\delta x \to 0} \frac{\delta y}{\delta x} = \lim_{\delta t \to 0} \frac{\delta y}{\delta t} \div \lim_{\delta t \to 0} \frac{\delta x}{\delta t}$

$$\frac{\mathrm{d}y}{\mathrm{d}x} = \frac{\mathrm{d}y}{\mathrm{d}x} \div \frac{\mathrm{d}x}{\mathrm{d}t}$$

> This follows from the arithmetic rule that dividing by a fraction is the same as multiplying by the reciprocal
> $$\frac{a}{c} \div \frac{b}{c} = \frac{a}{c} \times \frac{c}{b} = \frac{a}{b}.$$

> **KEY INFORMATION**
>
> If x and y are defined in terms of a parameter t, then
> $$\frac{\mathrm{d}y}{\mathrm{d}x} = \frac{\mathrm{d}y}{\mathrm{d}t} \div \frac{\mathrm{d}x}{\mathrm{d}t}.$$

Example 8

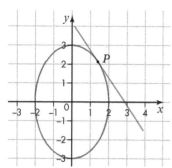

The equation of this ellipse is given parametrically as

$x = 2\sin t$ and $y = 3\cos t$ for $0 \leqslant t \leqslant 2\pi$.

A tangent has been drawn at the point P where $t = \frac{\pi}{4}$.

Find the equation of the tangent at P.

Solution

$x = 2 \sin t$

Differentiate x with respect to t.

$$\frac{dx}{dt} = 2 \cos t$$

$y = 3 \cos t$

Differentiate y with respect to t.

$$\frac{dy}{dt} = -3 \sin t$$

$$\frac{dy}{dx} = \frac{dy}{dt} \div \frac{dx}{dt}$$

> An alternative method is to eliminate the parameter and write y in terms of x.
>
> $\frac{dy}{dx}$ can be then be found.
>
> However this method is often more difficult.

$$= -3 \sin t \div 2 \cos t$$

$$= -\frac{3}{2} \frac{\sin t}{\cos t}$$

$$= -\frac{3}{2} \tan t$$

When $t = \frac{\pi}{4}$,

$$x = 2 \sin \frac{\pi}{4} = 2 \times \frac{1}{\sqrt{2}} = \frac{2}{\sqrt{2}}$$

$$y = 3 \cos t = 3 \cos \frac{\pi}{4} = 3 \times \frac{1}{\sqrt{2}} = \frac{3}{\sqrt{2}}$$

and

$$\frac{dy}{dx} = -\frac{3}{2} \tan \frac{\pi}{4} = -\frac{3}{2} \times 1 = -\frac{3}{2}.$$

The equation of the tangent at P is $y - \frac{3}{\sqrt{2}} = -\frac{3}{2}\left(x - \frac{2}{\sqrt{2}} \right)$.

Multiply by 2:

$$2y - \frac{3 \times 2}{\sqrt{2}} = -3x + \frac{3 \times 2}{\sqrt{2}}.$$

Rearrange.

$$2y + 3x = 6\sqrt{2}$$

> Use the fact that $\frac{2}{\sqrt{2}} = \sqrt{2}$.

Exercise 4.7A

1 The parametric equations of an ellipse are $x = 4 \cos t$ and $y = 3 \sin t$ for $0 \le t \le 2\pi$.

   a Find the coordinates of the point where $t = \frac{\pi}{4}$.

   b Show that the point $(0, -3)$ is on the ellipse.

   c Sketch the ellipse.

2 The parametric equation of part of a parabola is $x = t^2$ and $y = t + 2$ for $-3 \le t \le 3$.

   a Complete this table of values.

t	-3	-2	-1	0	1	2	3
x		4					
y		0					

   b Sketch the curve.

   c Show that $\frac{dy}{dx} = \frac{1}{2t}$.

   d Find the gradient of the curve at $(4, 0)$.

3 Find $\frac{dy}{dx}$ as a function of t for each of the curves given by the following parametric equations.

   a $x = t - 1$ and $y = t^2$

   b $x = t - 1$ and $y = t^4 - 2$

   c $x = t^2$ and $y = t^2 - 3t + 2$

PS 4 A curve is given by the parametric equations $x = t + 2$ and $y = t^2 - 1$. Show that the equation of the tangent to the curve when $t = 1$ is given by $2x - y - 6 = 0$.

PS 5 A curve is given by the parametric equations $x = \frac{t}{2}$ and $y = t^2 - 4$. Show that the equation of the normal to the curve when $t = -1$ is given by $8y - 2x + 23 = 0$.

6 Find $\frac{dy}{dx}$ as a function of t for each of the curves given by the following parametric equations.

   a $x = 4 \sin t$ and $y = 5 \cos t$

   b $x = 2 \cos t$ and $y = 3 \cos 2t$

   c $x = \tan 2t$ and $y = \sin 2t$

PS 7 A curve is given by the parametric equations $x = 3 \cos t + 1$ and $y = 7 \sin t - 4$. Show that when $t = \frac{\pi}{4}$ the exact value of the gradient is $-\frac{7}{3}$.

PS **8** The parametric equation of this curve is $x = 2\sin 2t$ and $y = 3\cos t$ for $\leqslant t \leqslant 2\pi$.

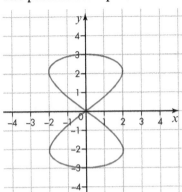

Find the gradient of each branch of the curve at the origin.

PS **9** A curve is given by the parametric equations $x = 2(t - \sin t)$ and $y = 2(1 - \cos t)$ for $0 \leqslant t \leqslant 2\pi$. Show that the gradients of the tangent and normal are $1 + \sqrt{2}$ and $1 - \sqrt{2}$, respectively, when $t = \dfrac{\pi}{4}$.

PS **10** A curve is given by the parametric equations $x = a\cos t$ and $y = b\sin t$ for $0 \leqslant t \leqslant 2\pi$.

In terms of a and b find the y-intercept of the tangent to the curve when $t = \dfrac{\pi}{4}$.

MM **11** This graph shows the shape of a speed bump designed to slow down traffic on a road. Lengths
PS are in centimetres.

The x-axis represents ground level.

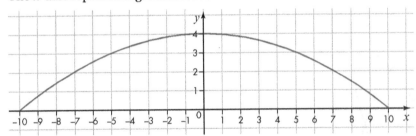

The curve of the speed bump is given parametrically by

$x = 10\sin t$ and $y = 2 + 2\cos 2t$ for $-\dfrac{\pi}{2} \leqslant t \leqslant \dfrac{\pi}{2}$.

a Find the gradient of the speed bump in terms of t.

b Find the value of t at the highest point on the speed bump.

c Find the angle of slope of the speed bump where $x = 9$, to the nearest degree.

12 The parametric equation of a curve is $x = \sin^2 \theta$, $y = 2\cos \theta$ for $0 \leqslant \theta < 2\pi$.

a Find $\dfrac{dy}{dx}$ in terms of θ.

b Find the equation of the tangent when the gradient is 1.

c Find the point where the tangent is parallel to the y-axis.

13 The parametric equation of a curve is $x = 10t$, $y = \dfrac{10}{t}$.

   a Find $\dfrac{dy}{dx}$ in terms of t.

   b Show that the equation of the tangent at $\left(10t, \dfrac{10}{t}\right)$ is $x + t^2 y = 20t$.

   c Find the points where the tangent meets the coordinate axes.

   d Show that the area of the triangle bounded by the tangent and the coordinate axes is constant.

14 The parametric equation of a curve is $x = t^2$, $y = t + \dfrac{1}{t}$.

   a Show that $\dfrac{dy}{dx} = \dfrac{t^2 - 1}{2t^3}$.

   b Find the stationary points on the curve.

   c Determine whether each stationary point is a maximum or a minimum.

15 The parametric equation of a curve is $x = t^2$, $y = 2t$.

   a Find $\dfrac{dy}{dx}$ in terms of t.

   b Find the equation of the normal to the curve at $(t^2, 2t)$.

   c Show that the area of the triangle formed by the normal and the coordinate axes is $\dfrac{1}{2}|t|(2 + t^2)^2$.

16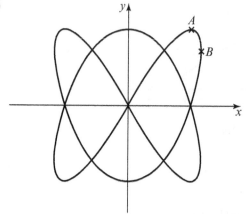

The parametric equation of this curve is $x = \sin 2t$, $y = \sin 3t$ for $0 \leqslant t < 2\pi$.

   a A is a maximum point in the first quadrant. Find its coordinates.

   b At point B the tangent is parallel to the y-axis. Find the coordinates of B.

   c Two branches of the curve pass through the origin. Find the gradient of each branch.

4.8 Differentiating implicit equations

This curve is an ellipse.

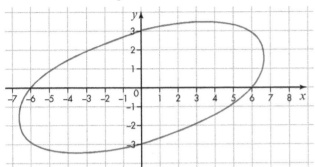

The equation is $x^2 + 4y^2 - 2xy = 36$.

This is an **implicit equation**. That means that it shows a relationship between coordinates of points on the curve but it is not written in an explicit form of the type $y = f(x)$.

| Stop and think | Why is it difficult to write the equation for y explicitly in this case? |

To find a point on the curve you can substitute a value for x (or for y) and then solve the resulting equation to find y (or x).

For example, if you substitute $x = 6$ in the equation it becomes

$36 + 4y^2 - 12y = 36$.

Rearrange.

$4y^2 - 12y = 0$

Divide by 4 and factorise.

$y^2 - 3y = 0$

$y(y - 3) = 0$

Therefore $y = 0$ or 3.

Two points on the curve are (6, 0) and (6, 3).

To find $\frac{dy}{dx}$, you can differentiate both sides of the equation.

$\frac{d}{dx}(x^2 + 4y^2 - 2xy) = \frac{d}{dx}(36)$

Differentiate each term on the left separately.

$\frac{d}{dx}(x^2) + \frac{d}{dx}(4y^2) - \frac{d}{dx}(2xy) = \frac{d}{dx}(36)$

The first and last terms are straightforward.

$2x + \frac{d}{dx}(4y^2) - \frac{d}{dx}(2xy) = 0$ ①

To differentiate $4y^2$, use the chain rule.

$$\frac{d}{dx}(4y^2) = \frac{d}{dy}(4y^2) \times \frac{dy}{dx}$$

$$= 8y\frac{dy}{dx}$$

To differentiate $2xy$, use the product rule.

$$\frac{d}{dx}(2xy) = 2\left\{\frac{d}{dx}(x) \times y + x\frac{dy}{dx}\right\}$$

$$= 2\left(y + x\frac{dy}{dx}\right)$$

Equation ① becomes

$$2x + 8y\frac{dy}{dx} - 2y - 2x\frac{dy}{dx} = 0. \qquad ②$$

Rearrange to make $\frac{dy}{dx}$ the subject.

$$8y\frac{dy}{dx} - 2x\frac{dy}{dx} = 2y - 2x$$

Divide by 2 and factorise.

$$4y\frac{dy}{dx} - x\frac{dy}{dx} = y - x$$

$$(4y - x)\frac{dy}{dx} = y - x$$

$$\frac{dy}{dx} = \frac{y - x}{4y - x} \qquad ③$$

> The expression for $\frac{dy}{dx}$ includes both x and y.

So the gradient of the curve at $(6, 3)$ is $\frac{3 - 6}{4 \times 3 - 6} = \frac{-3}{6} = -\frac{1}{2}$

and the gradient of the curve at $(6, 0)$ is $\frac{0 - 6}{4 \times 0 - 6} = \frac{-6}{-6} = 1$.

Stop and think   Look at the graph and check that result for the gradients are reasonable. Use the symmetry of the graph to write down the gradient at two other points.

You do not always need to rearrange equation ② to find $\frac{dy}{dx}$ in the explicit form of equation ③.

You could find the gradient at $(6, 3)$ by substituting the coordinates into equation ②

$$12 + 24\frac{dy}{dx} - 6 - 12\frac{dy}{dx} = 0$$

and solving that equation.

Example 9

Find the stationary points on the curve $x^2 + 4y^2 - 2xy = 36$.

Solution

This is the curve explored above.

You know that $2x + 8y\dfrac{dy}{dx} - 2y - 2x\dfrac{dy}{dx} = 0$.

At a stationary point $\dfrac{dy}{dx} = 0$.

Therefore $2x - 2y = 0$ so $x = y$.

Substitute x for y in the equation of the curve.

$x^2 + 4x^2 - 2x^2 = 36$

$\qquad 3x^2 = 36$

$\qquad x^2 = 12$

$\qquad x = \pm\sqrt{12}$

Since $x = y$, the stationary points are at $(\sqrt{12}, \sqrt{12})$ and $(-\sqrt{12}, -\sqrt{12})$.

Exercise 4.8A

1 Find an expression for $\dfrac{dy}{dx}$.

   a $x^2 - 2y^2 = 4$

   b $x(x + y) = 2y^2$

   c $\dfrac{1}{10} = \dfrac{1}{x^2} + \dfrac{1}{y^2}$

   d $\sin x \cos y = 0.5$

2 The equation of a curve is $2x^2 + y^2 = 34$.

   a Find an expression for $\dfrac{dy}{dx}$.

   b Find the gradient at $(3, 4)$.

   c Find the gradient at $(3, -4)$.

3 The equation of this curve is $x^2y = 4(2 - y)$.

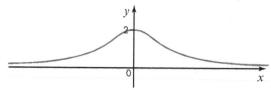

a Show that (2, 1) is on the curve.

b Find the gradient at (2, 1).

4 The equation of a curve is $x^2 + 4xy + y^2 = 25$.

a Show that (5, 0) and (0, 5) are on the curve.

b Find the gradient of the curve at (5, 0).

c Show that $\dfrac{dy}{dx} = -\dfrac{x + 2y}{2x + y}$.

(PS) 5 The equation of a curve is $y^2 - x^2 = 16$.

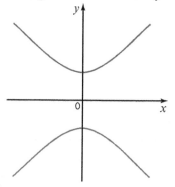

a Show that (3, 5) is on the curve.

b Find the equation of the tangent to the curve at (3, 5).

(PS) 6 The equation of this curve is $(x^2 + y^2)^2 = 50(x^2 - y^2)$.

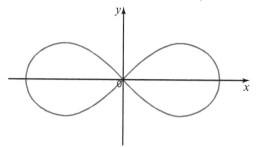

Show that stationary points lie on a circle, centre the origin, with a radius of 5.

PS **7** An implicit equation for this curve is $x^2 - y^2 = 1$.

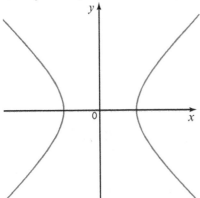

a Find an expression for $\dfrac{dy}{dx}$.

b The curve can be written parametrically as $x = \sec t$ and $y = \tan t$ for $0 \le t \le 2\pi$. Use this fact to find $\dfrac{dy}{dx}$ in terms of t.

c Show that your two expressions for $\dfrac{dy}{dx}$ are equivalent.

d Find the points on the curve where the gradient is equal to 2.

8 The equation of a curve is $x^2 + xy + y^2 = 13$.

a Show that $(-4, 3)$ is on the curve.

b Find the equation of the tangent to the curve at $(-4, 3)$.

9 The implicit equation of a curve is $x^2 + y^2 - 4x + 6y = 12$.

a Find the coordinates of the stationary points, A and B.

b Show that the point $P(-1, 1)$ is on the curve.

c Show that the normal at P passes through the midpoint of AB.

10 The equation of a curve is $x^2 + xy + y^2 = 7$.

a Show that $(3, -2)$ is on the curve.

b Find the equation of the normal at $(3, -2)$.

c Show that the normal meets the curve again.

SUMMARY OF KEY POINTS

> If $f(x) = e^x$, then $f'(x) = e^x$.

> If $f(x) = \ln x$, then $f'(x) = \dfrac{1}{x}$.

> If $f(x) = \sin x$, then $f'(x) = \cos x$.

> If $f(x) = \cos x$, then $f'(x) = -\sin x$.

> The product rule: if $f(x) = uv$ where u and v are functions of x, then $f'(x) = u\dfrac{dv}{dx} + v\dfrac{du}{dx}$.

> The quotient rule: if $f(x) = \dfrac{u}{v}$ where u and v are functions of x, then $f'(x) = \dfrac{v\dfrac{du}{dx} - u\dfrac{dv}{dx}}{v^2}$.

> If $f(x) = \tan x$, then $f'(x) = \sec^2 x$.

> If $f(x) = \tan^{-1} x$, then $f'(x) = \dfrac{1}{1 + x^2}$.

> If $x = f(t)$ and $y = g(t)$, then $\dfrac{dy}{dx} = \dfrac{dy}{dt} \div \dfrac{dx}{dt}$.

EXAM-STYLE QUESTIONS

1 A curve has equation $y = \sqrt{1 + x^2}$.

    a Find $\dfrac{dy}{dx}$.

    b Show that $\dfrac{dy}{dx} = \dfrac{x}{y}$.

2 Find the gradient of each of the following curves at the point at which $x = 0$.

    a $y = e^{-2x}$

    b $y = xe^{-x}$

3 For each of the following curves, find the exact gradient at the point indicated.

    a $y = x^2 \cos x$ at $\left(\dfrac{\pi}{2}, 0\right)$

    b $y = \dfrac{\cos x}{x^2}$ at $\left(\pi, -\dfrac{1}{\pi^2}\right)$

4 The equation of a curve is $y = \dfrac{x^n - 1}{x^n + 1}$.

    a In the case where $n = 2$, find an expression for the gradient of the curve.

    b In the case where n is an integer and $n \geqslant 2$, find an expression for the gradient of the curve.

5 A curve has equation $y = \ln x^2$ where $x > 0$. Find the coordinates of the point on the curve at which the gradient is 0.5.

6 The equation of a curve is $y = e^{2x} - 11e^x + 12x$. Find the exact x-coordinate of each of the stationary points of the curve and determine the nature of each stationary point.

7 The equation of a curve is $y = 0.3\sin 2x - 0.4\cos 2x$.

  a Find an expression for the gradient of the curve.

  b Show that $\dfrac{d^2y}{dx^2} + 4y = 0$.

8 The equation of this curve is $y = 0.1\,e^x + e^{-0.5x}$.

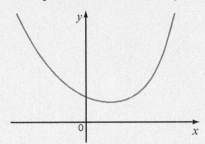

  a Find $\dfrac{dy}{dx}$.

  b The curve has a minimum point. Find its coordinates.

9 The parametric equations of a curve are:

$x = \theta - \sin\theta, \quad y = 1 - \cos\theta.$

  a Find an expression for the gradient of the curve in terms of θ.

  b In the case where the gradient of the curve is 0.5, show that $2\sin\theta + \cos\theta = 1$.

10 The equation of this graph is $y = e^{-x}\sin 10x$.

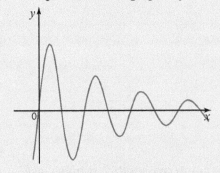

  a Obtain an expression for $\dfrac{dy}{dx}$.

A is the first maximum point with a positive x-coordinate. B is the first minimum point with a positive x-coordinate.

  b Show that the x-coordinate of A is 0.147 correct to 3 significant figures.

  c Find the x-coordinate of B, giving your answer correct to 3 significant figures.

11 The equation of a curve is $y = x^2 \tan 2x$ for $0 \leqslant x < \frac{\pi}{2}$. Find the gradient of the curve at the point for which $x = \frac{\pi}{6}$ and show that it can be written in the form $a\pi^2 + \frac{\pi}{\sqrt{b}}$, where a and b are constants.

12 The equation of a curve is $y = ae^{kx}$, where a and k are constants. The point $(10, 20)$ is on the curve and the tangent at that point passes through $(0, -5)$ on the y-axis. Find the values of a and k.

PS 13 A curve has equation $y = \sin^3 x$.

   a Show that $\dfrac{dy}{dx} = 3\cos x - 3\cos^3 x$.

   b Show that $\dfrac{d^2 y}{dx^2} = 6\sin x - 9\sin^3 x$.

14 The equation of this curve is $y = e^{-\frac{1}{2}x^2}$.

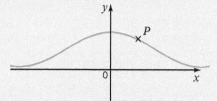

   a Show that the gradient of the curve at any point is given by the negative product of the coordinates at the particular point.

   b Given that the point P has a x-coordinate of a, show that the tangent at P crosses the x-axis at $\left(a + \dfrac{1}{a}, 0\right)$.

PS 15 The equation of this ellipse is $x^2 - 2xy + 2y^2 = 9$.

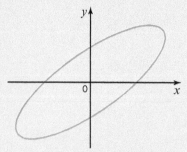

   a Find the coordinates of the maximum point.

   b Find the equation of the normal at the point where the graph crosses the positive x-axis.

C **16** It is given that $f(x) = \dfrac{x}{x+1}$ where $x \neq 1$.

    a Show that $f'(x) = \dfrac{1}{(x+1)^2}$.

The implicit equation of a curve is $\dfrac{x}{x+1} + \dfrac{y}{y+1} = xy^3$

    b Find the gradient of the curve at $(1, 1)$.

    c Find the equation of the normal to the curve at $(1, 1)$.

PS **17** The equation of a curve is $y = e^{-2x} \sin x$.

    a Obtain an expression for $\dfrac{dy}{dx}$ and show that it can be written in the form $ae^{-2x} \cos(x+b)$ where a and b are constants.

    b Find the gradient of the graph of curve at the point where it crosses the y-axis.

    c Find the coordinates of the stationary point on the curve that is closest to the y-axis.

18 The parametric equations of a curve are:

$x = 1 + 5\cos t, \quad y = 2 + 5\sin t.$

    a Find $\dfrac{dy}{dx}$ in terms of t.

    b Find the coordinates of the stationary points on the curve.

    c Find the equation of the normal at the point with parameter t.

    d Show that the normal passes through the point $(1, 2)$.

19 **a** Differentiate 2^x.

The equation of a curve is $y = a^{x^2}$, where a is a positive constant.

    b Show that $\dfrac{dy}{dx} = 2(\ln a)xy$.

The equation of a different curve is $y = x^x$, where $x > 0$.

    c Find $\dfrac{dy}{dx}$.

20 **a** Show that the derivative of $x \sec x$ is $\sec x(1 + x\tan x)$.

    b Differentiate $\ln\left(x + \sqrt{x^2+1}\right)$, simplifying your answer as much as possible.

PS **21** The parametric equations of a curve are:

$x = t^2 + 2, \quad y = 2t + 3.$

At point A on the curve the gradient is $-\dfrac{1}{2}$. At point B the gradient is 1.

    a Find the coordinates of A.

    b Show that the tangents at A and B intersect on the y-axis.

    c Find the area of the shape enclosed by the tangents at A and B and the x-axis.

22 A curve has equation $y = 2x \tan^{-1} x$ for $0 < x \leqslant \frac{\pi}{2}$.

    a Show that $\frac{dy}{dx}$ can be written in the form $\frac{f'(x)}{f(x)} + \frac{y}{x}$.

    b Find the equation of the normal to the curve at the point where $x = 1$.

23 A curve has equation $y = \tan^{-1}(e^x)$. Find the equation of the tangent to the curve at the point where $x = 0$.

Mathematics in life and work

This is a simple logistic curve, a shape that often occurs in practical situations to model data such as population growth and learning curves.

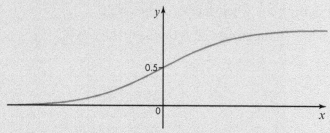

The equation of the curve is $y = \dfrac{e^x}{e^x + 1}$.

1 Find $\dfrac{dy}{dx}$.

2 Show that the equation satisfies the differential equation $\dfrac{dy}{dx} = y(1 - y)$.

3 A more complex equation is $y = \dfrac{e^{kx}}{e^{kx} + 1}$, where k is a positive constant.

    Find a formula for $\dfrac{dy}{dx}$ in terms of y in this case.

5 INTEGRATION

MATHEMATICS IN LIFE AND WORK

You know that differentiation and integration are inverse operations. Integration is often more difficult than differentiation – functions that are easy to differentiate may be difficult to integrate. That is why mathematicians have developed a range of special techniques to integrate particular types of function.

Integration techniques may be used in a variety of careers – for example:

> If you were an actuary, you could use integration when calculating fair premiums for life assurance policies.

> If you were a civil engineer, you could use integration to calculate the forces that might bend a steel beam used in construction.

> If you were an atomic physicist, you could use integration to find the velocity of particles in a particle accelerator.

> If you were a meteorologist, you could use integration to calculate average daily temperatures.

LEARNING OBJECTIVES

You will learn how to:

> use the trapezium rule to estimate a definite integral

> recognise integrals in particular forms

> use trigonometrical relationships in carrying out integration

> integrate using partial fractions

> integrate using a substitution

> use integration by parts.

LANGUAGE OF MATHEMATICS

Key words and phrases you will meet in this chapter:

> integration by parts, integrand, trapezium rule

PREREQUISITE KNOWLEDGE

You should already know how to:

> understand integration as the reverse of differentiation

> integrate $(ax + b)^n$ for any value of n except -1

> solve problems involving the evaluation of a constant of integration

> differentiate trigonometric and exponential functions

> differentiate using the product rule and the chain rule

> evaluate a definite integral to find the area under a curve.

You should be able to complete the following questions correctly:

1　Differentiate the following with respect to x.

   a $\sin(x^2 + 1)$ **b** $e^{-x}\cos 2x$ **c** $\ln(3x^2 + 2)$

2　Find:

   a $\displaystyle\int \frac{4}{\sqrt{x}}\,dx$ **b** $\displaystyle\int (2x - 1)^5\,dx$ **c** $\displaystyle\int \frac{4x^3 + 3}{x^2}\,dx$.

3　Find the area bounded by the curve $y = 10e^{-2x}$, the x-axis, the y-axis and the line $x = 1$.

5.1 The trapezium rule

This is a graph of $f(x) = e^{-0.5x^2}$, which is a function that occurs in statistics.

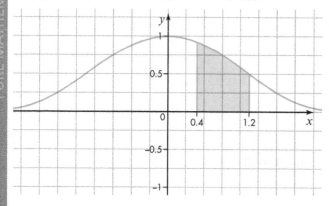

You often need to find an area such as the one shaded.

The area is $\displaystyle\int_{0.4}^{1.2} e^{-0.5x^2}\,dx$.

Unfortunately it is not possible to find an expression for the indefinite integral $\displaystyle\int e^{-0.5x^2}\,dx$.

An alternative is to estimate the value of the integral by drawing trapezia between the curve and the x-axis and calculating their areas.

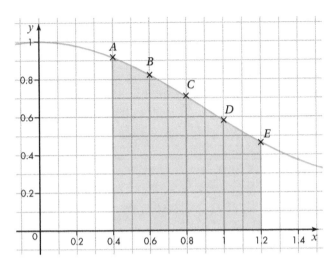

Four trapezia have been drawn by joining points A, B, C, D and E with straight line segments. The width of each trapezium is 0.2.

The coordinates of the points are included in this table.

Point	A	B	C	D	E
x	0.4	0.6	0.8	1.0	1.2
y	0.9231	0.8353	0.7261	0.6065	0.4868

Using the coordinates of A and B:

The area of the first trapezium is $0.2 \times \left(\dfrac{0.9231 + 0.8353}{2} \right) = 0.1758$.

Area of a trapezium $= \dfrac{a+b}{2} \times h$

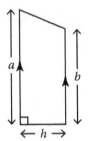

The total area of the four trapezia is

$$0.2 \times \left(\frac{0.9231 + 0.8353}{2} \right) + 0.2 \times \left(\frac{0.8353 + 0.7261}{2} \right)$$

$$+ \, 0.2 \times \left(\frac{0.7261 + 0.6065}{2} \right) + 0.2 \times \left(\frac{0.6065 + 0.4868}{2} \right)$$

$$= \frac{1}{2} \times 0.2 \{ 0.9231 + 2(0.8353 + 0.7261 + 0.6065) + 0.4868 \} = 0.5746.$$

Hence $\displaystyle \int_{0.4}^{1.2} e^{-0.5x^2} \, dx \approx 0.575$ to 3 d.p.

You can get a better estimate by using more trapezia.

In general, suppose the interval from a to b is divided into n equal intervals of width h, where $h = \dfrac{b-a}{n}$ and the y-coordinates are $y_0, y_1, y_2, \ldots y_{n-1}, y_n$. The y-values are known as ordinates. There is always one more ordinate than the number of strips.

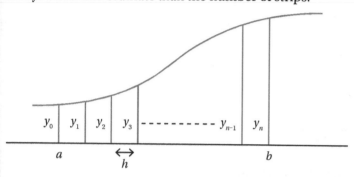

Then the sum of the areas of the trapezia is

$$\tfrac{1}{2}h(y_0 + y_1) + \tfrac{1}{2}h(y_1 + y_2) + \tfrac{1}{2}h(y_2 + y_3) + \ldots + \tfrac{1}{2}h(y_{n-1} + y_n)$$

$$= \tfrac{1}{2}h\{(y_0 + y_1) + (y_1 + y_2) + (y_2 + y_3) + \ldots + (y_{n-1} + y_n)\}$$

$$= \tfrac{1}{2}h\{y_0 + 2y_1 + 2y_2 + \ldots + 2y_{n-1} + y_n\}.$$

Hence $\displaystyle\int_a^b f(x)\,dx \approx \tfrac{1}{2}h\{y_0 + 2(y_1 + y_2 + \ldots + y_{n-1}) + y_n\}$.

This is called the **trapezium rule**.

Look at this graph of a different function.

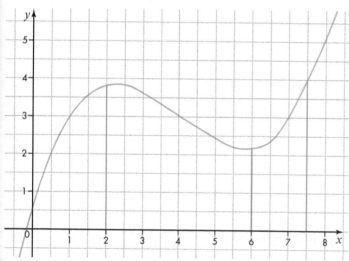

The trapezium rule will give an *underestimate* of $\displaystyle\int_0^2 f(x)\,dx$.

The trapezium rule will give an *overestimate* of $\displaystyle\int_6^{7.5} f(x)\,dx$.

> **KEY INFORMATION**
>
> The trapezium rule:
> $$\int_a^b f(x)\,dx \approx \tfrac{1}{2}h\{y_0 + 2(y_1 + y_2 + \ldots + y_{n-1}) + y_n\},$$
> where $h = \dfrac{b-a}{n}$.

Stop and think	Explain why the statements above are correct.

Will the trapezium rule estimate of $\int_{2}^{7.5} f(x)\,dx$ be an under-estimate or an over-estimate?

The next example shows how you can use the trapezium rule even if you do not know the equation of the curve.

Example 1

This table shows the speed of a car every five seconds in a 30-second interval.

Time (s)	0	5	10	15	20	25	30
Speed (ms⁻¹)	6.8	9.6	12.5	14.8	15.9	17.2	15.4

Use the trapezium rule to estimate the distance travelled in the 30-second interval.

Solution

The distance is the area under the speed–time graph, so you need to know $\int_{0}^{30} f(x)\,dx$.

$h = 5$ so the distance $\approx \dfrac{1}{2} \times 5\,\{6.8 + 2(9.6 + 12.5 + 14.8 + 15.9 + 17.2) + 15.4\}$

$= 405.5\,\text{m}.$

Exercise 5.1A

1 This table gives values of the function $f(x)$.

x	−2	0	2	4
$f(x)$	12.25	17.89	21.04	23.38

Use the trapezium rule to estimate $\int_{-2}^{4} f(x)\,dx$.

2 Here is a table for values of $\sqrt{1 - x^2}$.

x	0.2	0.3	0.4	0.5	0.6
$\sqrt{1 - x^2}$	0.9798	0.9539	0.9165	0.8660	0.8

Use the values in the table and the trapezium rule to estimate $\int_{0.2}^{0.6} \sqrt{1 - x^2}\,dx$.

Ⓒ Communication   ⓂⓂ Mathematical modelling   ⓅⓈ Problem solving

3 This is an ellipse.

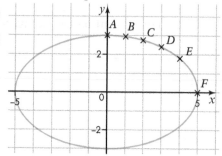

The coordinates of six points on the ellipse are included in this table.

Point	A	B	C	D	E	F
x	0	1	2	3	4	5
y	3	2.9394	2.7495	2.4	1.8	0

a Use the trapezium rule to estimate the area of the ellipse.

b Explain why your value is an underestimate.

4 a Sketch a graph of $y = 4e^{-x^2}$ for $0 \leq x \leq 2$.

b Use the trapezium rule with four strips to estimate the area bounded by the coordinate axes, the curve $y = 4e^{-x^2}$ and the line $x = 2$.

c Choose the correct statement from the following

A The trapezium rule will give an underestimate.

B The trapezium rule will give an overestimate.

C You cannot tell whether it is an underestimate or an overestimate.

5 a Use the trapezium rule with 10 intervals to estimate $\int_1^2 \frac{1}{x} \, dx$.

b Can you say whether the estimate is an underestimate or an overestimate?

6 a Sketch the graph of $y = 3x(4 - x)$.

b Find, by integration, the area between the curve and the x-axis.

c Explain why the trapezium rule will underestimate this area.

d Find the percentage error in using the trapezium rule with four intervals to estimate this area.

7 a Find an approximation to $\int_0^5 \sqrt{25 - x^2} \, dx$ by using the trapezium rule with 5 intervals.

b Show that the exact value of $\int_0^5 \sqrt{25 - x^2} \, dx$ is 6.25π.

8 The equations of two graphs are $y = \frac{1}{2}x^2$ and $y = 4\sqrt{x}$.

a Find the coordinates of the points where the graphs meet.

b Sketch the graphs of $\frac{1}{2} x^2$ and $y = 4\sqrt{x}$.

c Use the trapezium rule with four intervals to estimate the area between the curves.

d Find the exact area between the curves by integration.

9 $f(x) = xe^{-x}, x \geq 0$

    a Find the maximum value of $f(x)$.

    b Sketch the graph of $y = f(x)$ for $x \geq 0$.

    c Use a trapezium rule with four intervals to estimate $\int_0^2 xe^{-x} \, dx$.

10 **a** Sketch a graph of $y = 0.5^x, x \geq 0$.

    b Use the trapezium rule with three intervals to estimate $\int_0^3 0.5^x \, dx$.

    c Use a trapezium rule with 100 intervals to show that $\int_0^{100} 0.5^x \, dx \approx 1.5$.

Mathematics in life and work: Group discussion

As a meteorologist you would be interested in average temperatures. You can think of temperature as a function that varies with time.

The mean value of a function $f(x)$ over the interval (a, b) is defined as $\dfrac{\int_a^b f(x) \, dx}{b - a}$.

1 Suppose you have a record of the temperature on the hour, every hour from 06:00 to 18:00. How could you use the trapezium rule to estimate the average temperature?

2 In what way is your method in **Question 1** different from finding the mean of the 13 recorded temperatures? Would you get the same answer using each method?

3 A simple model of the temperature is:

If $f(x)$ is the temperature x hours after 06:00, then $f(x) = a + bx$ $(0 \leq x \leq 6)$ and $f(x) = a + b(12 - x)(6 < x \leq 12)$.

What does the graph of $f(x)$ look like? Can you calculate the average temperature?

5.2 Recognising integrals

Remember that integration is the inverse of differentiation.

What is $\int e^x \, dx$?

You know that if $f(x) = e^x$, then $f'(x) = e^x$.

Hence $\int e^x \, dx = e^x + c$.

What about $\int \sin x \, dx$?

If $f(x) = \cos x$, then $f'(x) = -\sin x$.

Hence $\int \sin x \, dx = -\cos x + c$.

Similarly, $\int \cos x \, dx = \sin x + c$.

In **Pure Mathematics 1 Chapter 7 Integration** you learnt how to find $\int (ax + b)^n \, dx$ where $n \neq -1$.

> **KEY INFORMATION**
>
> $\int e^x \, dx = e^x + c$
>
> $\int \sin ax \, dx = \dfrac{-1}{a} \cos ax + c$
>
> $\int \cos ax \, dx = \dfrac{1}{a} \sin ax + c$

For example, to find $\int 4\sqrt{2x+3}\,dx$, start with $f(x) = (2x+3)^{\frac{3}{2}}$ and then use the chain rule.

$$f'(x) = \frac{3}{2}(2x+3)^{\frac{1}{2}} \times 2 = 3\sqrt{2x+3}$$

You want the coefficient to be 4, so multiply by $f(x)$ by $\frac{4}{3}$.

$$\int 4\sqrt{2x+3}\,dx = \frac{4}{3}(2x+3)^{\frac{3}{2}} + c$$

You can use a similar method for integrals involving exponential or trigonometric functions.

Example 2

Find:

a $\int 2e^{4x+3}\,dx$ **b** $\int 2\sin(4x+3)\,dx$.

Solution

a You can guess that the integral is a multiple of e^{4x+3}.

If $f(x) = e^{4x+3}$ then, using the chain rule,

$$f'(x) = e^{4x+3} \times 4$$

$$= 4e^{4x+3}$$

You want the coefficient to be 2 and not 4, so multiply by $\frac{2}{4}$ or $\frac{1}{2}$.

$$\int 2e^{4x+3}\,dx = \frac{1}{2}e^{4x+3} + c$$

> You can check this is correct by differentiating. Remember always to include an arbitrary constant.

b In this case you might guess a multiple of $\cos(4x+3)$.

Differentiate using the chain rule again.

If $f(x) = \cos(4x+3)$, then

$$f'(x) = -\sin(4x+3) \times 4$$

$$= -4\sin(4x+3).$$

You want the coefficient to be 2, so multiply by $\frac{2}{-4} = -\frac{1}{2}$.

$$\int 2\sin(4x+3)\,dx = -\frac{1}{2}\cos(4x+3) + c$$

You know that $\int x^n\,dx = \frac{1}{n+1}x^{n+1} + c$, where $n \neq -1$. But what if $n = -1$?

What is $\int \frac{1}{x}\,dx$?

Suppose $\frac{dy}{dx} = \frac{1}{x}$.

Then $\frac{dx}{dy} = x$.

But you know that if $x = e^y$, then $\frac{dx}{dy} = e^y = x$.

So a solution is $x = e^y$.

Take logarithms.

$\ln x = y$

Therefore $\int \frac{1}{x} \, dx = \ln x + c$.

$\ln x$ is only defined if $x > 0$, so the result so far is

$\int \frac{1}{x} \, dx = \ln x + c$ where $x > 0$. ①

However, if $x < 0$ then $\ln(-x)$ exists. Using the chain rule:

$\frac{d}{dx} \ln(-x) = \frac{1}{-x} \times -1 = \frac{1}{x}$

This means that

$\int \frac{1}{x} \, dx = \ln(-x) + c$ where $x < 0$. ②

Combining results ① and ②, you can write:

$\int \frac{1}{x} \, dx = \ln|x| + c$ where $x \neq 0$.

> $\frac{dy}{dx} = \lim_{\delta x \to 0} \frac{\delta y}{\delta x}$ and
>
> $\frac{dx}{dy} = \lim_{\delta y \to 0} \frac{\delta x}{\delta y}$ so one is the
>
> reciprocal of the other.

> Remember that $|x|$ is the modulus of x. It is the value of x but ignoring any minus signs. For example, $|2.5| = |-2.5| = 2.5$.

Example 3

Find $\int \frac{1}{4x - 1} \, dx$.

Solution

In this case, you might guess a multiple of $\ln|4x - 1|$.

Using the chain rule:

$\frac{d}{dx} \ln|4x - 1| = \frac{1}{4x - 1} \times 4$

 $= \frac{4}{4x - 1}$.

Therefore $\int \frac{1}{4x - 1} \, dx = \frac{1}{4} \ln|4x - 1| + c$.

> **KEY INFORMATION**
>
> $\int \frac{1}{x} \, dx = \ln|x| + c$, where $x \neq 0$.

Exercise 5.2A

1 Find these integrals.

   a $\int \dfrac{1}{\sqrt{2x-3}}\,dx$ **b** $\int e^{2x-3}\,dx$ **c** $\int \sin(2x-3)\,dx$

2 Find these integrals.

   a $\int 4e^{-2x}\,dx$ **b** $\int (e^{0.5x} - e^{-0.5x})\,dx$ **c** $\int (2x + 3e^{4x+5})\,dx$

3 Integrate.

   a $\int (\sin 2x - \cos 2x)\,dx$ **b** $\int 4\cos(0.1x + 1.3)\,dx$ **c** $\int (4\sin 5x - 5\cos 4x)\,dx$

4 Find these integrals.

   a $\int \dfrac{5}{x}\,dx$ **b** $\int \dfrac{2}{5x}\,dx$ **c** $\int \dfrac{1}{5x+2}\,dx$

(PS) 5 **a** Find the value of $\int_0^\pi 2\cos 0.5x\,dx$.

   b Sketch a graph to show what area the definite integral represents.

(PS) 6 This is a graph of $y = \dfrac{4}{x}$.

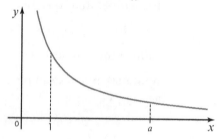

The region bounded by the curve, the x-axis and the lines $x = 1$ and $x = a$ has an area of 10.
Work out the value of a.

7 Work out these integrals.

   a $\int \dfrac{2}{x+1}\,dx$ **b** $\int \dfrac{3}{2x-1}\,dx$ **c** $\int \dfrac{x^2-6}{2x}\,dx$

8 This is a graph of $y = e^{0.5x-2}$.

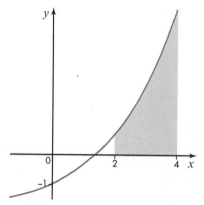

Calculate the area of the shaded region.

PS **9** **a** Differentiate $\sin(x^2)$.

  b Use your answer to **part a** to find $\int x\cos(x^2)\,dx$.

  c Work out $\int x\sin(x^2)\,dx$.

PS **10** Two students are asked to work out $\int \frac{1}{2x}\,dx$.
C Alice writes:

$\int \frac{1}{2x}\,dx = \frac{1}{2}\int \frac{1}{x}\,dx = \frac{1}{2}\ln x + c$.

Bhaskar writes:

Using the chain rule, $\frac{d}{dx}(\ln 2x) = \frac{1}{2x} \times 2 = \frac{1}{x}$

so $\frac{d}{dx}\left(\frac{1}{2}\ln 2x\right) = \frac{1}{2} \times \frac{1}{x} = \frac{1}{2x}$.

Therefore

$\int \frac{1}{2x}\,dx = \frac{1}{2}\ln 2x + c$.

Who is correct? Give a reason for your answer.

PS **11** **a** $f(x) = e^x (\sin x + \cos x)$.

  Find $f'(x)$.

  b Hence find $\int e^x \cos x\,dx$.

  c Find $\int e^x \sin x\,dx$.

12 $f(x) = x \ln x$

  a Find $f'(x)$.

  b Hence find $\int \ln x\,dx$.

  c If $\int_a^1 \ln x\,dx = 1$, find the value of a.

13 **a** Find $\int x\sqrt{x^2+1}\,dx$.

  b Find $\int \frac{x}{\sqrt[3]{x^2+1}}\,dx$.

14 **a** **i** Find $\int \frac{1}{(x-1)^3}\,dx$.   **ii** Hence find $\int \frac{x}{(x-1)^3}\,dx$.

  b **i** Find $\int \frac{1}{\sqrt{1+x}}\,dx$.   **ii** Hence find $\int \frac{x}{\sqrt{1+x}}\,dx$.

Mathematics in life and work: Group discussion

Engines in diesel vehicles have pistons moving up and down in cylinders.

The operation of a diesel engine can be illustrated by this graph.

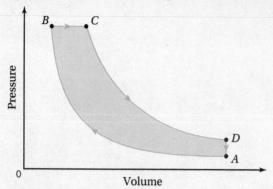

It shows how the pressure and volume of the fuel varies as the piston moves up and down.

The area enclosed by the loop measures the energy produced in each cycle.

As an automotive engineer, your job is to design an engine for which this area is as large as possible.

1   How can you calculate the area if you know the equations of the two curves?

2   Both the curves have an equation of the form $pv^\alpha = c$ where α and c are constants and $\alpha > 1$. Find a formula for the area under a curve of this type between $v = a$ and $v = b$.

5.3 Integration using trigonometrical relationships

In this section, you will look at integrals involving trigonometric functions.

First, remember that if $f(x) = \tan x$, then $f'(x) = \sec^2 x$.

Writing this in the form of an integral, you get
$\int \sec^2 x \, dx = \tan x + c$.

Example 4

Find $\int 5\sec^2(0.2x + 10) \, dx$.

Solution

A first guess is $f(x) = \tan(0.2x + 10)$.

Then $f'(x) = \sec^2(0.2x + 10) \times 0.2 = 0.2 \sec^2(0.2x + 10)$.

You want the coefficient to be 5 so multiply by $\dfrac{5}{0.2} = 25$.

Hence $\int 5\sec^2(0.2x + 10) \, dx = 25\tan(0.2x + 10) + c$.

You know that $\int \sin x \, dx = -\cos x + c$.

What about $\int \sin^2 x \, dx$?

You can find this by using the double angle formula
$\cos 2x = 1 - 2 \sin^2 x$.

Rearrange it.

$2 \sin^2 x = 1 - \cos 2x$

Hence

$\sin^2 x = \frac{1}{2} - \frac{1}{2} \cos 2x$

Now you can integrate.

$\int \sin^2 x \, dx = \int \left(\frac{1}{2} - \frac{1}{2} \cos 2x \right) dx$

$\qquad\qquad = \frac{1}{2}x - \frac{1}{4} \sin 2x + c$

In a similar way, you can find $\int \cos^2 x \, dx$ by using the formula
$\cos 2x = 2 \cos^2 x - 1$.

This gives the result $\int \cos^2 x \, dx = \frac{1}{2}x + \frac{1}{4} \sin 2x + c$.

> **Stop and think**   Check that the last result is correct.
> You know that $\sin^2 x + \cos^2 x \equiv 1$.
> Show that the formulae for $\int \sin^2 x \, dx$ and $\int \cos^2 x \, dx$ are consistent with this
> fact.

Exercise 5.3A

1 Find:

  a $\int \sec^2 x \, dx$ **b** $\int \sec^2(2x) \, dx$ **c** $\int \sec^2(2x - 1) \, dx$.

2 **a** Differentiate $\tan(x^2)$.

    b Hence find $\int x \sec^2(x^2) \, dx$.

 3 $\cot x \equiv \dfrac{\cos x}{\sin x}$

    a Use this identity to differentiate $\cot x$.

    b Hence find $\int \text{cosec}^2 x \, dx$.

4 a Show that $\int \cos^2 x \, dx = \frac{1}{2}x + \frac{1}{4}\sin 2x + c$.

  b Find $\int \cos^2 2x \, dx$.

5 a Differentiate $\ln|\cos x|$ with respect to x.

  b Use the result to **part a** to find $\int \tan x \, dx$.

  c Find $\int \cot x \, dx$.

6 Find these integrals.

  a $\displaystyle\int_0^{\frac{\pi}{2}} \sin x \cos x \, dx$ **b** $\displaystyle\int_0^{\frac{\pi}{8}} \sin 2x \cos 2x \, dx$

7 Work these out.

  a $\int \cos^2 5x \, dx$ **b** $\int \sin^2 0.5x \, dx$ **c** $\int \sin^2 ax \, dx$

PS 8

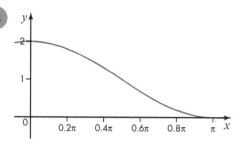

This is part of the graph of $y = 2\cos^2 0.5x$.

Find the area shown bounded by the curve and the positive coordinate axes.

9 Show that $\displaystyle\int_0^{\frac{\pi}{4}} \cos x \cos 3x \, dx = \int_0^{\frac{\pi}{4}} \sin x \sin 3x \, dx$.

10 Find these integrals.

  a $\int \sec^2(2x + 1) \, dx$ **b** $\int \mathrm{cosec}^2(3x + 2) \, dx$

5.4 Integrating $\dfrac{1}{x^2 + a^2}$

The starting point for this section is that if $f(x) = \tan^{-1} x$,

then $f'(x) = \dfrac{1}{x^2 + 1}$.

> This result was found in **Chapter 4 Differentiation.**

In integral form this is $\displaystyle\int \frac{1}{x^2 + 1} \, dx = \tan^{-1} x + c$.

Now look at $f(x) = \dfrac{1}{a}\tan^{-1}\dfrac{x}{a}$, where a is a constant.

Using the chain rule,

$$f'(x) = \frac{1}{a} \times \frac{1}{\left(\frac{x}{a}\right)^2 + 1} \times \frac{1}{a}$$

$$= \frac{1}{x^2 + a^2}.$$

Hence, in integral form, $\displaystyle\int \frac{1}{x^2 + a^2}\,dx = \frac{1}{a}\tan^{-1}\frac{x}{a} + c.$

> **KEY INFORMATION**
>
> $$\int \frac{1}{x^2 + a^2}\,dx = \frac{1}{a}\tan^{-1}\frac{x}{a} + c$$
>
> You need to remember this integral.

Example 5

Find:

a $\displaystyle\int \frac{1}{4x^2 + 1}\,dx$ **b** $\displaystyle\int \frac{2}{3x^2 + 4}\,dx.$

Solution

a You want the integrand in the form $\displaystyle\int \frac{1}{x^2 + a^2}\,dx.$

$$\int \frac{1}{4x^2 + 1}\,dx = \int \frac{1}{4(x^2 + 0.25)}\,dx$$

$$= \int \frac{0.25}{x^2 + 0.5^2}\,dx$$

$$= 0.25 \times \frac{1}{0.5}\tan^{-1}\frac{x}{0.5} + c$$

$$= \frac{1}{2}\tan^{-1} 2x + c$$

b $\displaystyle\int \frac{2}{3x^2 + 4}\,dx = \int \frac{2}{3\left(x^2 + \frac{4}{3}\right)}\,dx$

$$= \frac{2}{3}\int \frac{1}{x^2 + \left(\sqrt{\frac{4}{3}}\right)^2}\,dx$$

$$= \frac{2}{3} \times \sqrt{\frac{3}{4}}\,\tan^{-1}\sqrt{\frac{3}{4}}\,x + c$$

$$= \frac{2}{3} \times \frac{\sqrt{3}}{2}\,\tan^{-1}\frac{\sqrt{3}x}{2} + c$$

$$= \frac{1}{\sqrt{3}}\,\tan^{-1}\frac{\sqrt{3}x}{2} + c$$

The reciprocal of $\sqrt{\dfrac{4}{3}}$ is $\sqrt{\dfrac{3}{4}}$.

Exercise 5.4A

1 Work out these integrals.

a $\int \dfrac{1}{x^2 + 25}\,dx$

b $\int \dfrac{50}{x^2 + 100}\,dx$

c $\int \dfrac{6}{x^2 + 6}\,dx$

2 Work out these integrals.

a $\int \dfrac{10}{x^2 + 100}\,dx$

b $\int \dfrac{10}{100x^2 + 1}\,dx$

3 **a** Sketch a graph of $y = \dfrac{16}{x^2 + 4}$.

b Find the area bounded by the curve $y = \dfrac{16}{x^2 + 4}$, the x-axis and the lines $x = 4$ and $x = -4$.

4 The equation of a curve is $y = \dfrac{a}{x^2 + a^2}$.

Find the exact value of the area bounded by the curve, the positive coordinate axes and the line $x = a$.

5 Work out these integrals.

a $\int \dfrac{1}{5x^2 + 1}\,dx$

b $\int \dfrac{12}{5x^2 + 4}\,dx$

c $\int \dfrac{1}{px^2 + q}\,dx \quad (p > 0,\, q > 0)$

6 Work out these integrals.

a $\int \dfrac{x^2 + 1}{x^2}\,dx$

b $\int \dfrac{x^2}{x^2 + 1}\,dx$

7 Find $\int \dfrac{x}{x^4 + 1}\,dx$.

8 Work out these integrals.

a $\int \dfrac{10}{x^2 + 25}\,dx$

b $\int \dfrac{10x^2}{x^2 + 25}\,dx$

9 Find $\int \dfrac{(x + a)(x - a)}{x^2 + a^2}\,dx$.

10 **a** Sketch the graph of $y = \dfrac{10}{x^2 + 4}$.

b Find the area enclosed by the curve, the lines $x = 2$ and $x = -2$, and the x-axis.

c Find the area enclosed by the graph and the positive coordinate axes.

5.5 Integrating expressions of the form $\frac{f'(x)}{f(x)}$

You know that if $f(x) = \ln x$, then $f'(x) = \dfrac{1}{x}$.

In integral form, that is $\int \dfrac{1}{x}\,dx = \ln|x| + c$.

This leads to a method for integrating a wider group of functions.

Suppose you want to find $\int \dfrac{x^2}{x^3+1}\,\mathrm{d}x$.

Start with $f(x) = \ln\left|x^3+1\right|$.

Differentiate, using the chain rule.

$f'(x) = \dfrac{1}{x^3+1} \times 3x^2 = \dfrac{3x^2}{x^3+1}$

Notice that this is just $3 \times$ the integrand, so

$\int \dfrac{x^2}{x^3+1}\,\mathrm{d}x = \tfrac{1}{3}\ln\left|x^3+1\right| + c.$

This method works because the numerator is a multiple of the derivative of the denominator.

In general, $\int \dfrac{f'(x)}{f(x)}\,\mathrm{d}x = \ln|f(x)| + c.$

When you have to integrate a fraction, check whether the numerator is a multiple of the derivative of the denominator. The function you are integrating is called the **integrand**. In this example the integrand is $\dfrac{x^2}{x^2+1}$. If so, this method will give the answer.

> This is the logarithm of the denominator.

> **KEY INFORMATION**
>
> $\int \dfrac{f'(x)}{f(x)}\,\mathrm{d}x = \ln|f(x)| + c$

Example 6

Workout these integrals.

a $\displaystyle\int \dfrac{6}{x^2+9}\,\mathrm{d}x$ **b** $\displaystyle\int \dfrac{6x}{x^2+9}\,\mathrm{d}x$

Solution

a This involves the inverse tangent function.

$\displaystyle\int \dfrac{6}{x^2+9}\,\mathrm{d}x = 6\int \dfrac{1}{x^2+3^2}\,\mathrm{d}x$

$\qquad\qquad = 6 \times \dfrac{1}{3}\tan^{-1}\dfrac{x}{3} + c$

$\qquad\qquad = 2\tan^{-1}\dfrac{x}{3} + c$

b The derivative of x^2+9 is $2x$ and the numerator is a multiple of this.

If $f(x) = \ln\left|x^2+9\right|$, then $f'(x) = \dfrac{1}{x^2+9} \times 2x = \dfrac{2x}{x^2+9}$.

Hence $\displaystyle\int \dfrac{6x}{x^2+9}\,\mathrm{d}x = 3\ln\left|x^2+9\right| + c.$

> An alternative way of thinking about this is to rewrite the expression so that it contains the $\dfrac{f'(x)}{f(x)}$. In this case,
>
> $\displaystyle\int \dfrac{6x}{x^2+9}\,\mathrm{d}x$ can be rewritten
>
> as $3\displaystyle\int \dfrac{2x}{x^2+9}\,\mathrm{d}x.$
>
> Then $f(x) = 3\ln\left|x^2+9\right| + c.$

> The modulus signs are not strictly necessary in this particular example, since (x^2+9) is positive for all values of x.

Example 7

Find $\int \tan x \, dx$.

Solution

Write $\tan x = \dfrac{\sin x}{\cos x}$ and the integral becomes $\int \dfrac{\sin x}{\cos x} \, dx$.

But the derivative of $\cos x$ is $-\sin x$, which is the numerator $\times -1$ and hence

$$\int \frac{\sin x}{\cos x} \, dx = -\ln|\cos x| + c.$$

Exercise 5.5A

1 Find these integrals.

a $\displaystyle\int \frac{1}{x+1} dx$

b $\displaystyle\int \frac{x}{x^2+1} dx$

c $\dfrac{x^2}{x^3-3} dx$

2 Find these integrals.

a $\displaystyle\int \frac{x+1}{x^2+2x+5} dx$

b $\displaystyle\int \frac{x^2-x}{2x^3-3x^2+12} dx$

3 Find these integrals.

a $\displaystyle\int \frac{e^x+1}{e^x} dx$

b $\displaystyle\int \frac{e^x}{e^x+4} dx$

4 Find the area enclosed by the curve $y = \tan x$, the positive x-axis and the line $x = \dfrac{\pi}{4}$.

5 Find these integrals.

a $\displaystyle\int \tan 2x \, dx$

b $\displaystyle\int \cot 0.5x \, dx$

6 Show that $\displaystyle\int_0^a \frac{x}{x^2+a^2} dx = \ln\sqrt{2}$.

7 Find $\displaystyle\int \frac{1}{x\ln x} dx$, where $x > 1$.

8 Find these integrals.

a $\displaystyle\int_0^1 \frac{e^x+1}{e^x} dx$

b $\displaystyle\int_0^1 \frac{e^x}{e^x+1} dx$

c $\dfrac{1}{e^x+1}$

9 Integrate with respect to x.

a $\dfrac{\cos 2x}{\sin x \cos x + 4}$

b $\dfrac{\sec x}{\sin x - \cos x}$

10 $f(x) = \sec x$

a Show that $f'(x) = \sec x \tan x$. **b** Find $\displaystyle\int \frac{\tan x}{2+\cos x} dx$.

PURE MATHEMATICS 3

5.6 Integration using partial fractions

$\dfrac{x}{x^2+1}$, $\dfrac{1}{x^2+1}$ and $\dfrac{1}{x^2-1}$ are all examples of algebraic fractions.

How can you integrate each one?

1. $\displaystyle\int \dfrac{x}{x^2+1}\,dx$

   The derivative of x^2+1 is $2x$, which is double the numerator.

   So $\displaystyle\int \dfrac{x}{x^2+1}\,dx = \tfrac{1}{2}\ln\left|x^2+1\right| + c$ or simply $\tfrac{1}{2}\ln(x^2+1)+c$. •

   > The modulus sign is not needed because x^2+1 is positive for all values of x.

2. $\displaystyle\int \dfrac{1}{x^2+1}\,dx$

   You should recognise this as a particular example of the standard result.

   $\displaystyle\int \dfrac{1}{x^2+a^2}\,dx = \dfrac{1}{a}\tan^{-1}\dfrac{x}{a} + c$ with $a=1$.

   Hence $\displaystyle\int \dfrac{1}{x^2+1}\,dx = \tan^{-1}x + c$.

3. $\displaystyle\int \dfrac{1}{x^2-1}\,dx$

   This looks similar to the previous integral but the minus sign is an important difference.

   In this case, you can factorise the denominator: $x^2-1 = (x-1)(x+1)$. You can write the algebraic fraction in partial fractions, using the methods described in **Chapter 1 Algebra**.

   Suppose $\dfrac{1}{(x-1)(x+1)} \equiv \dfrac{A}{x-1} + \dfrac{B}{x+1}$.

   Hence $\dfrac{1}{(x-1)(x+1)} \equiv \dfrac{A(x+1)+B(x-1)}{(x-1)(x+1)}$.

   Equate the numerators.

   $1 \equiv A(x+1) + B(x-1)$

   Putting $x=1$, gives $1 = 2A$ or $A = \tfrac{1}{2}$.

   Putting $x=-1$, gives $1 = -2B$ or $B = -\tfrac{1}{2}$.

   So $\dfrac{1}{(x-1)(x+1)} \equiv \dfrac{\frac{1}{2}}{x-1} - \dfrac{\frac{1}{2}}{x+1}$ and you can rewrite the integral.

   $\displaystyle\int \dfrac{1}{(x-1)(x+1)}\,dx = \dfrac{1}{2}\int \dfrac{1}{x-1}\,dx - \dfrac{1}{2}\int \dfrac{1}{x+1}\,dx$

   $\qquad\qquad = \tfrac{1}{2}\ln|x-1| - \tfrac{1}{2}\ln|x+1| + c$

The solution could be written differently by using the laws of logarithms.

$$\int \frac{1}{(x-1)(x+1)}\,dx = \frac{1}{2}\ln\left|\frac{x-1}{x+1}\right| + c \text{ or } \ln\left|\frac{x-1}{x+1}\right|^{\frac{1}{2}} + c$$

Example 8

Find $\int \dfrac{5x^2 - 4x + 2}{(x-1)(x^2+2)}\,dx$.

Solution

You can write the integrand in partial fraction form.

$$\frac{5x^2 - 4x + 2}{(x-1)(x^2+2)} \equiv \frac{A}{x-1} + \frac{Bx+C}{x^2+2} \equiv \frac{A(x^2+2) + (Bx+C)(x-1)}{(x-1)(x^2+2)}$$

The second numerator is $Bx + C$ because the denominator is a quadratic.

Equate the numerators.

$$A(x^2+2) + (Bx+C)(x-1) \equiv 5x^2 - 4x + 2$$

If $x = 1$:

$$3A = 5 - 4 + 2$$
$$3A = 3$$
$$A = 1$$

Hence $x^2 + 2 + (Bx+C)(x-1) \equiv 5x^2 - 4x + 2$.

If $x = 0$:

$$2 - C = 2$$
$$C = 0$$

Hence $x^2 + 2 + Bx(x-1) \equiv 5x^2 - 4x + 2$.

If $x = 2$:

$$4 + 2 + 2B = 20 - 8 + 2$$
$$2B = 8$$
$$B = 4$$

Hence

$$\frac{5x^2 - 4x + 2}{(x-1)(x^2+2)} \equiv \frac{1}{x-1} + \frac{4x}{x^2+2}.$$

Then

$$\int \frac{5x^2 - 4x + 2}{(x-1)(x^2+2)}\,dx = \int \frac{1}{x-1}\,dx + \int \frac{4x}{x^2+2}\,dx$$

$$= \ln|x-1| + 2\ln\left|x^2 + 2\right| + c.$$

And this can be written as $\ln|x-1|(x^2+2)^2 + c$.

> **Stop and think** Look at the expression $\ln|x-1| + 2\ln|x^2+2| + c$ above.
>
> The modulus sign is needed on the first term, $\ln|x-1|$, but it is not needed on the second term, $\ln|x^2+2|$. Can you explain why?

Exercise 5.6A

1 **a** Write $\dfrac{x-10}{x(x+5)}$ as partial fractions.

   b Find $\displaystyle\int \dfrac{x-10}{x(x+5)}\,dx$.

2 Find these integrals.

   a $\displaystyle\int \dfrac{4}{x(x+4)}\,dx$
   b $\displaystyle\int \dfrac{1}{(x-2)(x-4)}\,dx$
   c $\displaystyle\int \dfrac{6}{x^2-9}\,dx$

3 Find $\displaystyle\int \dfrac{3x+4}{x(x+1)}\,dx$.

4 Find these integrals.

   a $\displaystyle\int \dfrac{4}{x^2-4}\,dx$
   b $\displaystyle\int \dfrac{4x}{x^2+4}\,dx$
   c $\displaystyle\int \dfrac{x^2}{x^2+4}\,dx$

5 Find $\displaystyle\int \dfrac{3x-6}{(x+2)(x-1)}\,dx$.

6 Find $\displaystyle\int \dfrac{4x+10}{(2x-3)(2x+1)}\,dx$.

7 Find $\displaystyle\int \dfrac{2x^2+9x-11}{(x+1)(x-2)(x+3)}\,dx$.

8 **a** Write $\dfrac{(x+1)(x+3)}{x(x^2+1)}$ in partial fractions.

   b Find $\displaystyle\int \dfrac{(x+1)(x+3)}{x(x^2+1)}\,dx$.

9 **a** Write $\dfrac{x^2-3x+8}{(x+2)(x-1)^2}$ in partial fractions.

   b Find $\displaystyle\int \dfrac{x^2-3x+8}{(x+2)(x-1)^2}\,dx$.

10 Find these integrals.

   a $\displaystyle\int \dfrac{x-1}{x^2-2x-8}\,dx$
   b $\displaystyle\int \dfrac{x-10}{x^2-2x-8}\,dx$

11 Find these integrals.

   a $\displaystyle\int \dfrac{x^2+3}{x^2+4}\,dx$
   b $\displaystyle\int \dfrac{x^2-3}{x^2-4}\,dx$

12 Show that $\displaystyle\int_2^5 \dfrac{x^2}{x^2-1}\,dx = 3 + \tfrac{1}{2}\ln 2$.

5.7 Integration using a substitution

Suppose you want to find $\int x\sqrt{2x+1}\,dx$.

Write this as $y = \int x\sqrt{2x+1}\,dx$ so that $\frac{dy}{dx} = x\sqrt{2x+1}$.

Make the substitution $u = 2x + 1$.

This can be rearranged as $x = \frac{1}{2}(u-1)$

and then $\frac{dy}{dx} = \frac{1}{2}(u-1)\sqrt{u}$.

You now have three variables, x, y, and u.

According to the chain rule, $\frac{dy}{du} = \frac{dy}{dx} \times \frac{dx}{du}$.

Hence $\frac{dy}{du} = \frac{1}{2}(u-1)\sqrt{u} \times \frac{dx}{du}$.

Looking at the definition $u = 2x + 1$ you can see that $\frac{du}{dx} = 2$, so

$\frac{dx}{du} = \frac{1}{2}$.

So $\frac{dy}{du} = \frac{1}{2}(u-1)\sqrt{u} \times \frac{1}{2} = \frac{1}{4}(u-1)\sqrt{u}$.

Integrate: $y = \int \frac{1}{4}(u-1)\sqrt{u}\,du$.

The integral with respect to x has been changed to an integral with respect to u; this is an integral you can find.

$$y = \frac{1}{4}\int \left(u^{\frac{3}{2}} - u^{\frac{1}{2}} \right) du$$

$$= \frac{1}{4}\left(\frac{2}{5}u^{\frac{5}{2}} - \frac{2}{3}u^{\frac{3}{2}} \right) + c$$

$$= \frac{1}{10}u^{\frac{5}{2}} - \frac{1}{6}u^{\frac{3}{2}} + c$$

You want the answer in terms of x, so substitute $2x + 1$ for u.

$$\int x\sqrt{2x+1}\,dx = \frac{1}{10}(2x+1)^{\frac{5}{2}} - \frac{1}{6}(2x+1)^{\frac{3}{2}} + c$$

Stop and think	Differentiate this answer to show that it is correct.

The general method is to change $\int f(x)\,dx$ to $\int g(u)\frac{dx}{du}\,du$.

A change of variable transforms f(x) into g(u).

KEY INFORMATION

Integration by substitution

$$\int f(x)\,dx = \int f(x)\frac{dx}{du}\,du$$

The following example shows how to set out the solution concisely.

Example 9

Find $\int \dfrac{1}{\sqrt{1-x^2}}\,dx$.

Solution

Make the substitution $x = \sin u$.

Then $\dfrac{dx}{du} = \cos u$.

Substitute.

$$\int \frac{1}{\sqrt{1-x^2}}\,dx = \int \frac{1}{\sqrt{1-\sin^2 u}} \times \frac{dx}{du}\,du$$

$$= \int \left(\frac{1}{\sqrt{1-\sin^2 u}} \times \cos u \right) du$$

$$= \int \left(\frac{1}{\cos u} \times \cos u \right) du$$

$$= \int 1\,du$$

$$= u + c$$

Finally, as $x = \sin u$, $u = \sin^{-1} x$

So $\int \dfrac{1}{\sqrt{1-x^2}}\,dx = \sin^{-1} x + c$.

> Notice that in this case, x is written in terms of u rather than the other way round.

> In this case, all of the terms have cancelled.

The substitution is not an obvious one to make, but it works because it eliminates the square root in the denominator.

Knowing what to choose for u becomes easier with practice. Your choice may not always work. Be prepared to try again with something different.

There are no definitive rules about how to choose the substitution, which is one reason why integration is harder than differentiation.

Stop and think Here is an integral: $\int x(x^2 + 2)^2\,dx$.

You might be able to guess the form of the answer, as in **Section 5.2**.

Alternatively you could use the substitution $u = x^2 + 2$.

Do both methods give the same answer?

PURE MATHEMATICS 3

Exercise 5.7A

1 Find $\int x(x+2)^3 \, dx$. Use the substitution $u = x + 2$.

2 **a** Find $\int \dfrac{x}{\sqrt{x^2+4}} \, dx$. Use the substitution $u = x^2 + 4$.

   b Find $\int \dfrac{x}{\sqrt{x+4}} \, dx$. Use the substitution $u = x + 4$.

3 Work out $\int x^2(2x^3+5)^4 \, dx$. Use the substitution $u = 2x^3 + 5$.

4 **a** Use the substitution $u = x^2$ to find $\int x \sin x^2 \, dx$.

   b Find $\int_0^{\frac{\pi}{2}} x \sin x^2 \, dx$.

5 **a** Work out $\int \dfrac{1}{x+2} \, dx$.

   b Use the substitution $u = x + 2$ to find $\int \dfrac{x}{x+2} \, dx$.

   c Use the substitution $u = x + 2$ to find $\int \dfrac{x^2}{x+2} \, dx$.

6 Work these out.

   a $\int x\sqrt{x^2+3} \, dx$ **b** $\int x\sqrt{x+3} \, dx$ **c** $\int x^2\sqrt{x+3} \, dx$

7 **a** Use the substitution $u = \cos x$ to show that $\int \tan x \, dx = -\ln|\cos x| + c$.

   b Use a substitution to show $\int_{\frac{\pi}{4}}^{\frac{\pi}{2}} \cot x \, dx = \dfrac{1}{2} \ln 2$.

8 Use the substitution $x = a \sin u$ to find $\int \dfrac{1}{\sqrt{a^2-x^2}} \, dx$.

9 The diagram shows part of the graph of $y = x(x-2)^4$.

   a Find the coordinates of the point where the curve touches the x-axis.

   b Find the area between the curve and the x-axis.

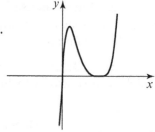

10 The diagram shows part of the graph of $y = x\sqrt{x+4}$.

   Find the area between the graph and the negative x-axis.

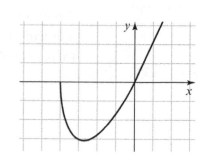

11 Work out these integrals.

   a $\int \sin\theta\sqrt{1-\cos\theta}\,d\theta$
   b $\int \sin\theta\sqrt{1-\cos^2 2\theta}\,d\theta$
   c $\int \sin\theta\cos 2\theta\,d\theta$

12 Work out these integrals.

   a $\int \dfrac{1}{\sqrt{1+x}}\,dx$
   b $\int \dfrac{x}{\sqrt{1+x}}\,dx$
   c $\int_0^1 \dfrac{x^3}{\sqrt{1+x^2}}\,dx$

5.8 Integration by parts

Look at this integration:

$$\int x\sin x\,dx$$

This is the product of two different functions. Although you can easily integrate each function separately, this does not help you to integrate the product.

> You cannot just integrate each function separately and multiply the results. The integral is *not* $\frac{1}{2}x^2 \times (-\cos x)$.

The product rule for differentiation provides a useful tool for integrations of this type.

$$\frac{d}{dx}(uv) = u\frac{dv}{dx} + v\frac{du}{dx}$$

Write this in an integral form.

$$uv = \int u\frac{dv}{dx}\,dx + \int v\frac{du}{dx}\,dx$$

Rearrange.

$$\int u\frac{dv}{dx}\,dx = uv - \int v\frac{du}{dx}\,dx$$

This may not look very useful, but it can transform an integral into something more manageable.

Look again at $\int x\sin x\,dx$ and compare it with $\int u\frac{dv}{dx}\,dx$.

You are going to make two substitutions.

They are $u = x$ and $\frac{dv}{dx} = \sin x$.

Differentiate one and integrate the other.

$$\frac{du}{dx} = 1 \text{ and } v = -\cos x$$

> Don't worry about including an arbitrary constant yet.

Now use the formula $\int u\frac{dv}{dx}\,dx = uv - \int v\frac{du}{dx}\,dx$.

$$\int x\sin x\,dx = x \times -\cos x - \int -\cos x \times 1\,dx$$
$$= -x\cos x + \int \cos x\,dx$$

PURE MATHEMATICS 3

The integration is now straightforward.

$$\int x \sin x \, dx = -x \cos x + \sin x + c$$

This method is called **integration by parts**.

Use integration by parts with a product, equating one term to u and the other to $\frac{dv}{dx}$.

Choosing $u = x$ gave $\frac{du}{dx} = 1$ and that made the integration simpler.

> Differentiate $-x \cos x + \sin x + c$ to show that it is correct.

> It is important to get these the right way round.

Stop and think   Show that choosing $u = \sin x$ and $\frac{dv}{dx} = x$ makes the integration more difficult.

Example 9

Find $\int x^2 e^{2x} \, dx$.

Solution

Make the substitutions $u = x^2$ and $\frac{dv}{dx} = e^{2x}$.

Then $\frac{du}{dx} = 2x$ and $v = \frac{1}{2} e^{2x}$.

$$\int u \frac{dv}{dx} \, dx = uv - \int v \frac{du}{dx} \, dx$$

So

$$\int x^2 e^{2x} \, dx = x^2 \times \frac{1}{2} e^{2x} - \int \frac{1}{2} e^{2x} \times 2x \, dx$$

$$\int x^2 e^{2x} \, dx = \frac{1}{2} x^2 e^{2x} - \int x e^{2x} \, dx. \qquad ①$$

To find $\int x e^{2x} \, dx$, use integration by parts again.

Make the substitutions $u = x$ and $\frac{dv}{dx} = e^{2x}$.

Then $\frac{du}{dx} = 1$ and $v = \frac{1}{2} e^{2x}$.

So

$$\int x e^{2x} \, dx = x \times \frac{1}{2} e^{2x} - \int \frac{1}{2} e^{2x} \, dx$$

$$= \frac{1}{2} x e^{2x} - \frac{1}{4} e^{2x} + c.$$

Substitute into ①.

$$\int x^2 e^{2x} \, dx = \frac{1}{2} x^2 e^{2x} - \frac{1}{2} x e^{2x} + \frac{1}{4} e^{2x} + c$$

KEY INFORMATION

Integration by parts

$$\int u \frac{dv}{dx} \, dx = uv - \int v \frac{du}{dx} \, dx$$

Stop and think   Check that the answer is correct by differentiating.

At first glance, the following example may not look like a question that involves integration by parts.

Example 10

Find $\int \ln x \, dx$.

Solution

Make the substitutions $u = \ln x$ and $\frac{dv}{dx} = 1$.

Then $\frac{du}{dx} = \frac{1}{x}$ and $v = x$.

$$\int u \frac{dv}{dx} \, dx = uv - \int v \frac{du}{dx} \, dx$$

$$\int \ln x \, dx = x \ln x - \int \left(x \times \frac{1}{x} \right) dx$$

$$= x \ln x - \int 1 \, dx$$

$$= x \ln x - x + c$$

Stop and think   Check that this result is correct by differentiating.

Choosing which part of a product to differentiate and which to integrate can be difficult.

This table gives guidance about which term to choose as u to differentiate, in various cases.

Type of product	Examples	What to choose as u
A power of x and a trigonometric term	$x \cos 2x$	The power of x
	$x^2 \sin x$	x or x^2
	$x \tan^{-1} x$	
A power of x and a power of e	$x e^{2x}$	The power of x
	$x^2 e^{-x}$	x or x^2
A power of x and a logarithmic term	$x \ln x$	The logarithmic term
	$x^2 \ln 2x$	$\ln x$ or $\ln 2x$
A power of e and a trigonometric term	$e^x \sin x$	Either term
	$e^{-x} \cos 2x$	

Stop and think   If one of the terms of the product is a power of x, it is often that term which should be differentiated. Why does choosing that term as $\frac{dv}{dx}$ to integrate usually make the problem more difficult?

Are there any exceptions to this rule?

PURE MATHEMATICS 3

Exercise 5.8A

1 Work out these integrals.

   a $\int x\cos x\,dx$ **b** $\int (x+1)\cos x\,dx$

2 Work out these integrals.

   a $\int x\sin 2x\,dx$ **b** $\int x\cos 4x\,dx$

3 Work out these integrals.

   a $\int x\,e^{x}\,dx$ **b** $\int x\,e^{-x}\,dx$ **c** $\int x\,e^{0.5x}\,dx$

4 Find $\int_{0}^{1}x^{2}e^{x}\,dx$.

5 Find $\int (2x^{2}-1)e^{-x}\,dx$.

6 Find $\int_{0}^{\frac{\pi}{2}}e^{x}\cos x\,dx$.

7 Work out these integrals.

   a $\int_{1}^{e}\ln x\,dx$ **b** $\int_{1}^{e}\ln x^{2}\,dx$ **c** $\int_{1}^{e}\ln 2x\,dx$ **d** $\int_{1}^{e}\ln(x+1)\,dx$

8 **a** Find $\int e^{-x}\sin 2x\,dx$.

   b Here is a graph of $y=e^{-x}\sin 2x$, where $x\geqslant 0$.

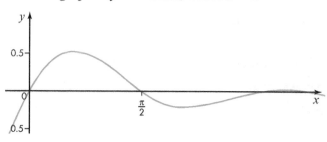

   Find the area bounded by the curve and the x-axis from 0 to $\frac{\pi}{2}$.

(PS) 9 **a** Use a substitution to find $\int x\sqrt{x+1}\,dx$.

   b Use integration by parts to find $\int x\sqrt{x+1}\,dx$.

   c Show that your answers to **parts a** and **b** are the same.

10 By first expressing $\cos^{3}x=\cos^{2}x\times\cos x$, find $\int\cos^{3}x\,dx$.

11 Find $\int_{0}^{\frac{\pi}{2}}x^{2}\sin x\,dx$.

12 **a** Show that $f(x)=xe^{-x}$ has a maximum value when $x=1$.

   b Find the area bounded by the curve $y=xe^{-x}$, the x-axis and the line $x=1$.

SUMMARY OF KEY POINTS

> $\int x^n \, dx = \frac{1}{n+1} x^{n+1} + c, \, n \neq -1$

> $\int e^{ax} \, dx = \frac{1}{a} e^{ax} + c$

> $\int \sin ax \, dx = -\frac{1}{a} \cos ax + c$

> $\int \cos ax \, dx = \frac{1}{a} \sin ax + c$

> $\int \frac{1}{x} \, dx = \ln|x| + c, \, x \neq 0$

> $\int \frac{1}{x^2 + a^2} \, dx = \frac{1}{a} \tan^{-1} \frac{x}{a} + c$

> $\int \frac{f'(x)}{f(x)} = \ln|f(x)| + c$

> Integration by substitution: $\int f(x) \, dx = \int f(x) \frac{dx}{du} \, du$

> Integration by parts $\int u \frac{dv}{dx} \, dx = uv - \int v \frac{du}{dx} \, dx$

EXAM-STYLE QUESTIONS

1 **a** Use the trapezium rule with four intervals to find an approximation to

$\int_1^{1.8} e^{2x+1} \, dx$.

b Find the exact value of

$\int_1^{1.8} e^{2x+1} \, dx$.

2 Use the trapezium rule with five intervals to find an approximation to

$\int_1^2 \frac{x}{x^3 + 2} \, dx$.

3 **a** Find $\int \sin(4x + 1) \, dx$.

b Find the exact value of $\int_2^3 \frac{1}{4x+1} \, dx$.

PS **4** **a** Find $\int \sec^2 0.5x \, dx$.

b Hence find $\int \tan^2 0.5x \, dx$.

5 The graph shows the region bounded by the lines $y = e^{0.5x}$, $y = e^{-0.5x}$ and $x = 3$.

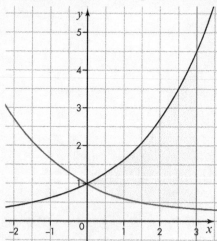

Find the area of the region.

6 **a** Find $\int \sin^2 2x \, dx$.

b

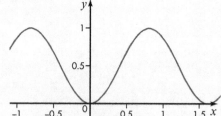

This is a graph of $y = \sin^2 2x$.

Find the area of the region bounded by the curve, the x-axis and the line $x = 1$.

7 **a** Find $\int \sqrt{2x+1} \, dx$.

b Use the substitution $u = 2x + 1$ to find $\int 2x\sqrt{2x+1} \, dx$.

c

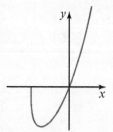

The graph shows the curve $y = 2x\sqrt{2x+1}$.

Find the area of the region bounded by the curve below the x-axis.

8 **a** Find $\int (3\cos 3x - \tan^2 x) \, dx$.

b Find the exact value of $\int_2^{2.5} 2e^{4x-8} \, dx$.

9 **a** Prove that $\cos^2 2x - 4\sin^2 x \cos^2 x \equiv \cos 4x$.

   b Hence find the exact value of $\int_0^{\frac{\pi}{16}} (\cos^2 2x - 4\sin^2 x \cos^2 x)\,dx$.

10 **a** Find $\int \dfrac{2+x}{x^2}\,dx$.

    b Prove that $9\cos^2\theta - \sin^2\theta - 6\sin\theta\cos\theta \equiv 5\cos 2\theta - 3\sin 2\theta + 4$.

    c Hence find $\int(9\cos^2\theta - \sin^2\theta - 6\sin\theta\cos\theta)\,dx$.

11 Work out each integral.

    a $\displaystyle\int \dfrac{1}{x^2+9}\,dx$

    b $\displaystyle\int \dfrac{x}{x^2+9}\,dx$

    c $\displaystyle\int \dfrac{x^2}{x^2+9}\,dx$

12 **a** Use the substitution $u = \cos 0.5x$ to find $\int \tan 0.5x\,dx$.

    b Show that $\int_0^{\frac{\pi}{2}} \tan 0.5x\,dx = \ln 2$.

13 $f(x) = x\sin 0.5x$.

    a Show that at a stationary point $\tan 0.5x + 0.5x = 0$.

    b Sketch the graph of $y = f(x)$ for $0 \leqslant x \leqslant 2\pi$.

    c Find $\int x\sin 0.5x\,dx$.

    d Find the area of between the curve $y = f(x)$ and the x-axis when $0 \leqslant x \leqslant 2\pi$.

14 **a** Find $\int_0^{\frac{\pi}{2}} \sin x\cos x\,dx$.

    b Find $\int x\sin x\cos x\,dx$.

15 **a** Write $\dfrac{x+7}{x^2-x-6}$ in partial fractions.

    b Find $\displaystyle\int \dfrac{x+7}{x^2-x-6}\,dx$.

16 **a** Find $\displaystyle\int \dfrac{x-1}{x^2-2x-3}\,dx$.

    b Express $\dfrac{4x}{x^2-2x-3}$ in partial fractions.

    c Show that $\displaystyle\int_0^1 \dfrac{4x}{x^2-2x-3}\,dx = \ln\left(\dfrac{16}{27}\right)$.

17 Let $f(x) = x^2 e^{-x}$ $x > 0$.

    a Find the maximum value of $f(x)$.

    b Find $\int f(x)\,dx$.

18 **a** Find $\int_0^1 \sqrt{1-x}\,dx$.

    b Find $\int_0^1 x\sqrt{1-x}\,dx$.

    c Find $\int_0^1 x^2\sqrt{1-x}\,dx$.

19 Work out each integral.

    a $\int \dfrac{8}{1+4x}\,dx$

    b $\int \dfrac{8}{1+4x^2}\,dx$

    c $\int \dfrac{8}{1-4x^2}\,dx$

20 **a** Find the exact value of $\int_0^{\frac{\pi}{2}} \sin x \cos x\,dx$.

    b Find the exact value of $\int_0^{\frac{\pi}{2}} \sin^2 x \cos x\,dx$.

    c Use the trapezium rule with 6 intervals to show that $\int_0^{\frac{\pi}{2}} \sin^2 x \cos^2 x\,dx \approx \dfrac{\pi}{16}$.

    d Show by integration that $\dfrac{\pi}{16}$ is the exact value of the integral in **part c**.

Mathematics in life and work

You are investigating the work done when a piston in an engine cylinder moves out.

You have the following values for the volume and the pressure, in suitable units.

Volume, v	5	10	15	20	25
Pressure, p	5.80	2.52	1.55	1.10	0.84

The work done is $\int_5^{25} p\,dv$.

1 Use the trapezium rule to estimate the work done.

2 From theoretical considerations you think that the pressure and volume are connected by the formula $pv^{1.2} = 40$. Use this formula to calculate the work done.

Under different conditions the formula connecting p and v is $pv = 24$.

3 Is more or less work done in this case? Give a reason for your answer.

6 NUMERICAL SOLUTION OF EQUATIONS

MATHEMATICS IN LIFE AND WORK

This chapter deals with numerical methods that start with an approximation to a solution and then use iteration (repeated processes) to reach an approximation closer to the actual solution. The mathematics used here is widely applicable in a range of careers – for example:

> If you were a computer programmer, you might need to use a step-by-step process called an algorithm to solve a problem. Numerical methods can be used in algorithms.

> If you were a statistician, you might want to analyse population growth. Numerical methods can be used to approximate solutions to equations that cannot be solved by usual methods.

> If you were a mathematician, you might want to find solutions to mathematical problems without relying on calculators or computers. Numerical methods can be employed to find approximate solutions to a specified degree of accuracy.

In this chapter, you will consider the application of numerical methods in computer programming.

LEARNING OBJECTIVES

You will learn how to:

> locate, approximately, a root of an equation, by means of graphical considerations and/or searching for a sign change

> understand the idea of, and use the notation for, a sequence of approximations which converges to a root of an equation

> understand how a given simple iterative formula relates to the equation being solved, and use a given iteration to determine a root to a prescribed degree of accuracy.

LANGUAGE OF MATHEMATICS

Key words and phrases you will meet in this chapter:

> cobweb diagram, convergent sequence, divergent sequence, iteration, iterative formula, root, staircase diagram, sub-interval

PREREQUISITE KNOWLEDGE

You should already know how to:

> ❯ simplify and manipulate algebraic expressions
> ❯ work with coordinates in all four quadrants
> ❯ identify and interpret roots (solutions) of equations
> ❯ use and understand the graphs of functions
> ❯ sketch curves defined by simple equations including quadratics, cubics, quartics and reciprocals
> ❯ interpret the algebraic solution of equations graphically.

You should be able to complete the following questions correctly:

1 Sketch the graph of $y = x^2 - 4x - 5$.

2 Sketch the graph of $y = x^3 - x^2 - 2x$.

3 Sketch the graph of $y = \dfrac{-3}{x+1}$, clearly indicating any asymptotes.

4 Sketch the graph of $y = -\dfrac{2}{(2x-1)^2}$, clearly indicating any asymptotes.

6.1 Finding roots

In **Pure Mathematics 1 Chapter 1 Quadratics** you solved quadratic equations to find their **roots** (solutions). The method was to ensure that $f(x) = 0$ and then, using factorisation, the quadratic formula or completing the square, solve the equation to find the root(s).

For example, find the roots of $f(x) = x^2 + x - 6$.

$x^2 + x - 6 = 0$

$(x - 2)(x + 3) = 0$

$x = -3$ or $x = 2$

So $f(x) = x^2 + x - 6$ has solutions $x = -3$ or $x = 2$ when $f(x) = 0$, that is, when the curve cuts the x-axis. If you hadn't been asked to solve $f(x) = x^2 + x - 6$ but instead to show that $f(x)$ has a root between -4 and -2, how could you have done this?

Here is a graph of this function.

What is significant about the value of $f(x)$ between -4 and the root (which you know to be at -3)?

$f(x) > 0$ because the curve is above the x-axis, so all the y values are positive.

Similarly, what is significant about the value of $f(x)$ between the root (which you know to be at -3) and -2?

$f(x) < 0$ because the curve is below the x-axis, so all the y values are negative.

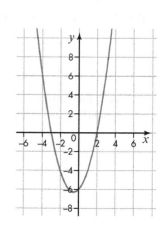

If you can demonstrate that there is a sign change in f(x) between –4 and –2, then you have shown that there is a root between these values.

$f(x) = x^2 + x - 6$

$f(-4) = 16 - 4 - 6 = 6$

$f(-2) = 4 - 2 - 6 = -4$

There is a sign change between f(–4) and f(–2); therefore there is a root of f(x) between –4 and –2.

Watch out, though! There are some instances, which will be discussed in **Section 6.2**, that are exceptions to this rule.

KEY INFORMATION

In general, if there is an interval in which the sign of the continuous function f(x) changes, then the interval contains a root of f(x) = 0.

Example 1

a Draw the graph of $y = x^2 - 7x + 5$ for $-1 \leqslant x \leqslant 7$. How many solutions are there to this equation in this interval? You must justify your answer.

b Show that $x^2 - 7x + 5 = 0$ has a root in the interval $0 < x < 1$.

Solution

a Draw a table of values for $y = x^2 - 7x + 5$.

x	−1	0	1	2	3	4	5	6	7
y	13	5	−1	−5	−7	−7	−5	−1	5

Plot the points on a pair of axes and join with a smooth curve.

The curve crosses the x-axis twice in the interval $-1 \leqslant x \leqslant 7$. Consequently, there are two solutions to the equation $y = x^2 - 7x + 5$ in this interval.

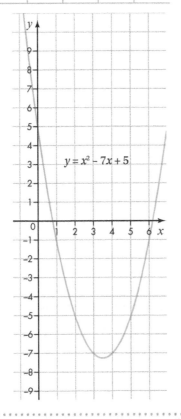

If the question had asked you to sketch the graph, then an alternative method would have been to complete the square. You could then sketch the graph by considering it as a transformation of the curve $y = x^2$.

b $f(x) = x^2 - 7x + 5$

$f(0) = 0 - 0 + 5 = 5$

$f(1) = 1 - 7 + 5 = -1$

There is a sign change between f(0) and f(1). Therefore, there is a root of f(x) in the interval $0 < x < 1$.

If there hadn't been a sign change when the interval limits were substituted into the function then you would have shown that $x^2 - 7x + 5 = 0$ did not have a root in the interval $0 < x < 1$ or that it had two roots in this interval.

Stop and think   Would you have identified a root if the interval given was $-1 \leqslant x \leqslant 7$? How does the interval relate to the identification of roots?

Example 2

Show that $\sin x = 0$ has a root in the interval $\frac{\pi}{2} < x < \frac{3\pi}{2}$.

Note that the interval has been given in radians so you need to make sure that your calculator is in radian mode.

Solution

$f(x) = \sin x$

$f\left(\frac{\pi}{2}\right) = 1$

$f\left(\frac{3\pi}{2}\right) = -1$

There is a sign change between $f\left(\frac{\pi}{2}\right)$ and $f\left(\frac{3\pi}{2}\right)$; therefore there is a root of $f(x)$ in the interval $\left(\frac{\pi}{2}\right) < x < \left(\frac{3\pi}{2}\right)$.

Remember that trigonometric functions have an infinite number of roots, so make sure that intervals are narrow enough to identify a sign change.

Example 3

a   On the same set of axes, sketch the graphs of $y = x^3$ and $y = x^2 + 2x$ for $-3 \leqslant x \leqslant 3$. Use your graphs to show that the equation $x^3 = x^2 + 2x$ has three roots.

b   Show that one root of the equation $x^3 = x^2 + 2x$ lies in the interval $1.8 < x < 2.1$.

Solution

a   You need to sketch the graphs of $y = x^3$ and $y = x^2 + 2x$ on the same pair of axes.

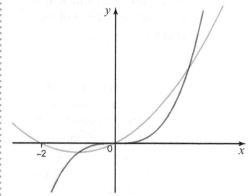

Highlight the three points of intersection on the graph as these are the roots of the equation $x^3 = x^2 + 2x$.

b   Rearrange the equation into the form $f(x) = 0$.

$x^3 - x^2 - 2x = 0$

$f(x) = x^3 - x^2 - 2x$

Substitute $x = 1.8$ (the lower bound of the given interval) into $f(x)$.

$f(1.8) = 5.832 - 3.24 - 3.6 = -1.008$

Substitute $x = 2.1$ (the upper bound of the given interval) into $f(x)$

$f(2.1) = 9.261 - 4.41 - 4.2 = 0.651$

There is a sign change between $f(1.8)$ and $f(2.1)$.

Therefore, there is a root of $x^3 = x^2 + 2x$ in the interval $1.8 < x < 2.1$.

Exercise 6.1A

1 Show that each of the following equations has a root in the given interval.

    a $3x^2 + 4x - 11 = 0$ $1.2 < x < 1.4$

    b $x^3 + 6x^2 + 11x + 6 = 0$ $-1.4 < x < -0.8$

    c $8x^4 + 2x^3 - 53x^2 + 37x - 6 = 0$ $1 < x < 3$

2 Given that $f(x) = 1 - e^x - \ln x$, show that the equation $f(x) = 0$ has a root in the interval $0.2 < x < 0.7$.

3 Show that $f(x) = \dfrac{\sin x}{e^x}$ does not have a root in the interval $-0.8 < x < -0.7$ when $f(x) = 0$.

4 **a**   On the same pair of axes, sketch the graphs of $y = \dfrac{1}{x^2}$ and $y = e^x$, where $-4 \leqslant x \leqslant 4$.

        Use your graphs to show that the equation $\dfrac{1}{x^2} = e^x$ has one root.

    b   Show that the root of the equation $\dfrac{1}{x^2} - e^x = 0$ lies in the interval $0.6 < x < 0.8$.

Ⓒ 5 Does the equation $\sin^2 x - \cos^2 x = 0$, where $0 < x < 2\pi$, have a root near to $x = \dfrac{\pi}{4}$? You must show and explain all your mathematical reasoning.

Ⓟˢ Ⓒ 6 **a**   On the same pair of axes, sketch the graphs of $y = 7 + \dfrac{5}{x}$ and $y = \dfrac{13}{x^2}$, where $-6 \leqslant x \leqslant 6$.

        Using your graph, write down the number of roots for the equation $7 + \dfrac{5}{x} = \dfrac{13}{x^2}$.

    b   Show that a root of the equation $7 + \dfrac{5}{x} = \dfrac{13}{x^2}$ lies in the interval $0.95 < x < 1.1$.

    c   Explain how solving the equation $7x^2 + 5x - 13 = 0$ gives the solutions to $7 + \dfrac{5}{x} = \dfrac{13}{x^2}$.

        Hence find the roots of $7 + \dfrac{5}{x} = \dfrac{13}{x^2}$ to 3 s.f.

7 Show algebraically that the equation $x = 3x^2 - \dfrac{1}{x}$ has a root in the interval $0 < x < 1$.

8 Show algebraically that the equation $x = 3 - \dfrac{1}{x} - \dfrac{2}{x^2}$ has three roots in the intervals $-1 < x < 0$, $1.4 < x < 1.8$ and $1.8 < x < 2.2$.

(C) **9** A student says that the equation $e^x = \frac{1}{2x} + 7$ has one root in the interval $0 < x < \frac{3}{2}$. Explain their mistake. You must justify your answer.

(PS) **10** The equation $2^x = 3x^2 - 1$ has three roots in the interval $-5 < x < 10$. By evaluating different intervals, determine a distinct interval for each root.

6.2 How change of sign methods can fail

In **Example 1**, you saw that $x^2 - 7x + 5 = 0$ has two roots: one in the interval $0 < x < 1$ and the other in the interval $6 < x < 7$. Would these roots have been identified if the interval had been $-1 \leqslant x \leqslant 7$?

$f(x) = x^2 - 7x + 5$

From the table of values in **Example 1**:

$f(-1) = 13$

$f(7) = 5$

There is no sign change between $f(-1)$ and $f(7)$ but this does *not* mean that there are no roots of $f(x)$ between -1 and 7. There isn't a sign change because there are two roots. At the first root the sign changes from positive to negative and on the second it changes from negative to positive. The interval used is too wide and needs to be narrowed.

> **KEY INFORMATION**
>
> If there are an even number of roots for $f(x) = 0$ in the given interval, then a sign change will not be detected.

Example 4

This is a sketch of $f(x) = 5 - 7x - 3x^2$. By evaluating intervals, show that $f(x) = 0$ has one root in each of the intervals $-3 < x < -2$ and $0 < x < 1$.

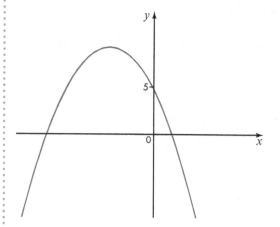

Solution

$f(x) = 5 - 7x - 3x^2$

$f(-3) = 5 + 21 - 27 = -1$

$f(-2) = 5 + 14 - 12 = 7$

There is a sign change between f(–3) and f(–2). Therefore, there is a root of f(x) in the interval $-3 < x < -2$.

f(0) = 5

f(1) = 5 – 7 – 3 = –5

There is a sign change between f(0) and f(1); therefore there is a root of f(x) in the interval $0 < x < 1$.

There are cases when f(x) changes sign but there is not a root.

Consider $f(x) = \frac{1}{x}$ in the interval $-1 < x < 1$.

$f(-1) = \frac{1}{-1} = -1$

$f(1) = \frac{1}{1} = 1$

There is a sign change between f(–1) and f(1) but there isn't a root.

Here is the graph of $f(x) = \frac{1}{x}$.

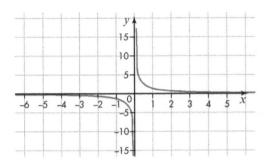

In the interval $-1 < x < 1$, there is a sign change but there isn't a root. Instead, there is a discontinuity at $x = 0$. The line $x = 0$ is an asymptote.

Example 5

Show that f(x) = tan x does not have a root in the interval $\frac{\pi}{4} < x < \frac{3\pi}{4}$.

Solution

f(x) = tan x

$f\left(\frac{\pi}{4}\right) = 1$

$f\left(\frac{3\pi}{4}\right) = -1$

There is a sign change between $f\left(\frac{\pi}{4}\right)$ and $f\left(\frac{3\pi}{4}\right)$.

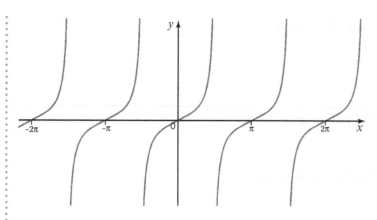

There is a discontinuity at $x = \frac{\pi}{2}$, so although there is a sign

change between $f\left(\frac{\pi}{4}\right)$ and $f\left(\frac{3\pi}{4}\right)$, there isn't a root.

Exercise 6.2A

1 This is a sketch of $f(x) = 6x^2 - x - 2$.
By evaluating the intervals
$-2 < x < 0$ and $0 < x < 2$,
show that $f(x) = 0$ has two roots.

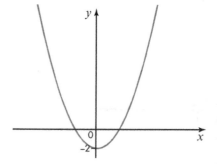

2 By evaluating intervals, show that the function $f(x) = \frac{1}{(x - 3)}$ has a discontinuity in the interval $2.4 < x < 3.3$.

C 3 Explain why evaluation of roots cannot be used to show that $f(x) = (x \pm a)^2$ has a root in the interval $a - 1 < x < a + 1$.

4 The function $f(x) = \frac{1}{x} + 2$ has a root and an asymptote in the interval $-1 < x < 1$. Sketch the function on a pair of axes. Given that a sub-interval lies within a given interval:

a use a sub-interval to show a sign change for the root

b use a sub-interval to show a sign change for the asymptote.

PS 5 For $f(x) = \tan x$, where $\frac{\pi}{4} < x < \frac{5\pi}{4}$, choose suitable intervals to demonstrate the sign change of the function for each root and each asymptote.

6 This is a sketch of $f(x) = \dfrac{1}{(3x+2)(2x-1)}$.

Find intervals, to one decimal place, for each of the discontinuities and show that the sign of f(x) changes within the intervals.

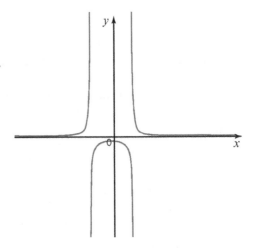

7 **a** Sketch the graph of $f(x) = \dfrac{1}{x^2} - 2$.

 b Show whether the function $f(x) = \dfrac{1}{x^2} - 2$ has a sign change in each of the intervals $-\dfrac{1}{2} < x < \dfrac{1}{2}$, $-1 < x < -\dfrac{1}{2}$ and $\dfrac{1}{2} < x < 1$.

 c Explain the cause where there is a sign change in the intervals in **part b**.

8 The function $f(x) = \dfrac{1}{x+4} - 5$ has two sign changes in the interval $-6 < x < -2$. Find suitable sub-intervals for each sign change. Show that a root is the cause of only one of the sign changes.

9 By evaluating sub-intervals, show that the equation $e^{x+1} - 7 = x^3$ has two solutions in the interval $-2 < x < 2$.

10 The function $f(x) = x^4 - 5x^2 + 1$ has 4 roots in the interval $-4 < x < 4$. Find the sub-interval for each root.

Mathematics in life and work: Group discussion

Two computer programmers are discussing the algorithms they have programmed to identify the number of roots for any equation.

1 Programmer A says that their algorithm detects sign changes between consecutive integer values. Is this an appropriate method to identify all the possible roots? If not, why not?

2 Programmer B says that their algorithm is much more sensitive and always tries to find one negative and one positive root. Is this an appropriate method to identify all the possible roots? If not, why not?

3 If you had to write an algorithm to identify the number of roots for any equation, what are the key items that you need to consider?

6.3 Iterative methods

An **iteration** is the repetition of a process or action. You can use iteration to find a sequence of approximations, each one getting closer to a root of $f(x) = 0$. You can find an approximation in the sequence using a specified approach, then use the result of this approximation in the next iteration until you have the required level of accuracy.

To solve an equation in the form $f(x) = 0$ by an iterative method, you first need to rearrange $f(x) = 0$ into a form where $x = g(x)$. Then use the **iterative formula** $x_{n+1} = g(x_n)$ for the iterations, where x_n is the latest value from the formula and x_{n+1} is the next value. There may be more than one rearrangement of $f(x) = 0$.

$x_{n+1} = g(x_n)$ converges to a root, a, provided that the initial approximation x_0 is close to a and $-1 < g'(a) < 1$, where $g'(a)$ is the value of the first derivative of $g(x_n)$ when $x_n = a$.

If each iteration gets closer to the root, the sequence is **convergent.**

KEY INFORMATION

x_n is the latest value from the iterative formula and x_{n+1} is the next value.

KEY INFORMATION

To solve an equation in the form $f(x) = 0$ by an iterative method, rearrange $f(x) = 0$ into a form where $x = g(x)$. Subsequently, use the formula $x_{n+1} = g(x_n)$ for the iterations.

Example 6

a Show that one root of the equation $x^2 = 7x + 3$ lies in the interval $7 < x < 8$.

b Show that $x^2 = 7x + 3$ can be written in the form
$x = 7 + \dfrac{3}{x}$.

c Find the root of $x^2 = 7x + 3$ in the interval $7 < x < 8$ to 3 d.p.

d Show graphically the first two iterations for this formula.

Solution

a Rearrange the equation so that $f(x) = 0$.

$x^2 - 7x - 3 = 0$

$f(7) = 49 - 49 - 3 = -3$

$f(8) = 64 - 56 - 3 = 5$

There is a sign change between $f(7)$ and $f(8)$; therefore there is a root of $x^2 = 7x + 3$ in the interval $7 < x < 8$.

b $x^2 = 7x + 3$

Divide both sides of the equation by x.

$x = 7 + \dfrac{3}{x}$

c Specify the iterative formula.

As $x = 7 + \dfrac{3}{x}$,

the iterative formula will be $x_{n+1} = 7 + \dfrac{3}{x_n}$.

Although an iteration method is being used here, this question could be solved exactly using the quadratic formula.

Let $x_0 = 7$ (the lower bound of the interval containing the root).

$$x_1 = 7 + \frac{3}{x_0}$$

$$= 7 + \frac{3}{7}$$

$$= 7.428\,571\,429$$

$$x_2 = 7 + \frac{3}{x_1}$$

$$= 7 + \frac{3}{7.428\,571\,429}$$

$$= 7.403\,846\,154$$

$$x_3 = 7.405\,194\,805$$

$$x_4 = 7.405\,121\,010$$

The last two iterations gave the same answer to 3 d.p., so you can now stop.

So a root of $x^2 = 7x + 3$ in the interval $7 < x < 8$ to 3 d.p. is 7.405.

d To show graphically the iterations for the formula, you first need to draw the graphs of $y = x$ and $y = 7 + \frac{3}{x}$ on the same pair of axes. The graphs intersect when $x = 7 + \frac{3}{x}$, which is the iterative formula you have used.

> You need to be able to use the memory function on your calculator to store and recall the value from each iteration.

> For completeness, the outcome of each iteration in some examples is accurate to 9 decimal places, but in general it is acceptable to give intermediate working accurate to 6 decimal places'

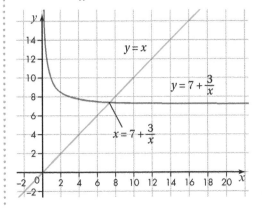

If you substitute $x = 7$ into $y = 7 + \dfrac{3}{x}$ then $y = 7.428\,571\,429$. On the graph this is the same as moving vertically from $x = 7$ to $y = 7.428\,571\,429$.

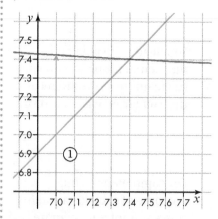

Now let $x = 7.428\,571\,429$. This is a horizontal movement to $y = x$.

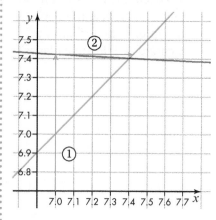

Substitute $x = 7.428\,571\,429$ into $y = 7 + \dfrac{3}{x}$ then $y = 7.403\,846\,154$.

On the graph this is the same as moving vertically from $x = 7.428\,571\,429$ to $y = 7.403\,846\,154$.

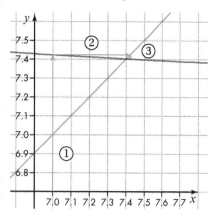

Now let $x = 7.403\,846\,154$, which is a horizontal movement to $y = x$.

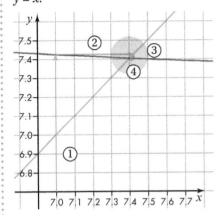

You have produced a **cobweb diagram.** A more general form is shown below.

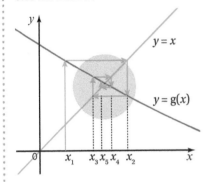

Stop and think   In **Example 6b** the function was rearranged into a specific iterative formula. Are there other ways in which this function could have been rearranged? Generally, can there be more than one iterative formula for $f(x) = 0$?

Example 7

a   Show that $x^3 - 5x^2 - 4x + 7 = 0$ can be written in the form $x = a + \dfrac{b}{x} + \dfrac{c}{x^2}$. State the values of constants a, b and c.

b   Hence, or otherwise, use a suitable iterative formula to find an approximate solution of the equation to 2 decimal places with $x_0 = 1$.

c   Show graphically the first two iterations for this formula.

Solution

a $x^3 - 5x^2 - 4x + 7 = 0$

Divide all terms in the equation by x^2.

$$x - 5 - \frac{4}{x} + \frac{7}{x^2} = 0$$

Rearrange the equation to make x the subject.

$$x = 5 + \frac{4}{x} - \frac{7}{x^2}$$

State the values of a, b and c.

$a = 5$, $b = 4$, $c = -7$

b Specify the iterative formula.

As $x = 5 + \frac{4}{x} - \frac{7}{x^2}$,

the iterative formula will be $x_{n+1} = 5 + \frac{4}{x_n} - \frac{7}{(x_n)^2}$.

$x_0 = 1$

$$x_1 = 5 + \frac{4}{1} - \frac{7}{1} = 2$$

$x_2 = 5.25$

$x_3 = 5.507\,937$

$x_4 = 5.495\,486$

$x_5 = 5.496\,085$

So a root of $x^3 - 5x^2 - 4x + 7 = 0$ with $x_0 = 1$ to 2 d.p. is 5.50.

c To show graphically the iterations for the formula, you first need to draw the graphs of $y = x$ and $y = 5 + \frac{4}{x} - \frac{7}{x^2}$ on the same pair of axes.

The graphs intersect when $x = 5 + \frac{4}{x} - \frac{7}{x^2}$, which is the iterative formula you have used.

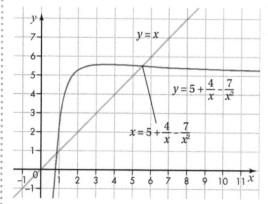

If you substitute $x = 1$ into $y = 5 + \dfrac{4}{x} - \dfrac{7}{x^2}$ then $y = 2$. On the

graph this is the same as moving vertically from $x = 1$ to $y = 2$.

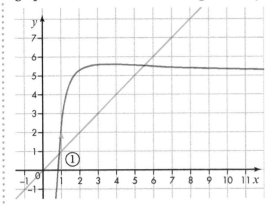

Now let $x = 2$. This is a horizontal movement to $y = x$.

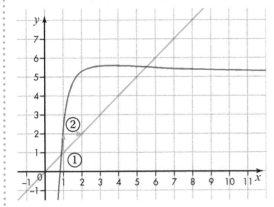

Substitute $x = 2$ into $y = 5 + \dfrac{4}{x} - \dfrac{7}{x^2}$; then $y = 5.25$. On the graph

this is the same as moving vertically from $x = 2$ to $y = 5.25$.

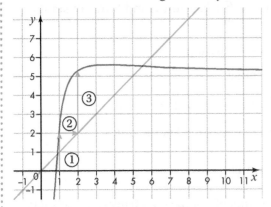

Now let $x = 5.25$. This is a horizontal movement to $y = x$.

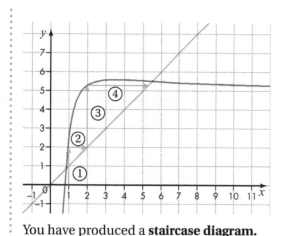

You have produced a **staircase diagram.**

What is the relationship between g(x), the iterative formula and the type of graphical representation? Is it possible to have a combined diagram? If so, under what circumstances?

Exercise 6.3A

1 Show that $x^2 = 11 - 5x$ can be written in each of the following forms.

   a $x = \sqrt{11 - 5x}$ **b** $x = \dfrac{11}{x} - 5$ **c** $x = \dfrac{x^2 - 11}{-5}$

2 The equation $x^3 - 6 = x^2$ has a root in the interval $2 < x < 3$. Use the iteration formula

   $x_{n+1} = \sqrt{\left(\dfrac{6}{x_n - 1}\right)}$, starting with $x_0 = 2$, to find x_3 to 3 d.p.

3 **a** Show that one root of the equation $x^2 + 4x - 7 = 0$ lies in the interval $-6 < x < -5$.

   b Show that $x^2 + 4x - 7 = 0$ can be written in the form $x = \dfrac{a}{x} + b$, where a and b are constants to be found.

   c Find the root of $x^2 + 4x - 7 = 0$ in the interval $-6 < x < -5$ to 2 d.p.

   d Show graphically the first two iterations for this formula.

PS 4 **a** Show that the equation $2x^3 - 5x + 1 = 0$ has a root in the interval $-2 < x < -1$.

   b Using a suitable iterative formula, find the root of $2x^3 - 5x + 1 = 0$ in the interval $-2 < x < -1$ to 3 d.p.

   c Show graphically the first two iterations for this formula.

PS 5 **a** Using the iterative formula $x_{n+1} = \left(7 - 3x_n^2\right)^{\frac{1}{5}}$, find the root of $x^5 + 3x^2 - 7 = 0$ near $x = 1$ to 2 d.p.

   b Show graphically the first two iterations for this formula.

(PS) 6 Using the iterative formulae $x_{n+1} = \dfrac{x_n^3 + 1}{4}$ and $x_{n+1} = (4x_n - 1)^{\frac{1}{3}}$, find the two positive roots of the equation $x^4 - 4x^2 + x = 0$ to 2 d.p.

7 The function $f(x) = x^3 - x^2 - 3x + 3$ has three roots in the interval $-2 < x < 2$. Find the root to which the iterative formula $x_{n+1} = \dfrac{x_n^3 - x_n^2 + 3}{3}$ with $x_0 = 1.5$ converges.

(C) 8 **a** Show that the equation $x^2 - 3x - 13 = 0$ has a root in each of the intervals $-3 < x < -2$ and $5 < x < 6$.

   b Compare the merits of each of the following iterative formulae to find the root in the interval $5 < x < 6$.

   i $x_{n+1} = \dfrac{x_n^2 - 13}{3}$ **ii** $x_{n+1} = \left(3x_n + 13\right)^{\frac{1}{2}}$

9 Using the iterative formula $x_{n+1} = \left(x_n^2 + 4\right)^{\frac{1}{4}}$, find the root of $x^4 - x^2 - 4 = 0$ in the interval $1 < x < 2$.

10 Using the iterative formula $x_{n+1} = \left(\dfrac{1}{2}\left(x_n^2 + 8\right)\right)^{\frac{1}{5}}$, find the root of $2x^5 - x^2 - 8 = 0$ in the interval $1 < x < 2$.

6.4 How iterative methods can vary

As you have seen, equations can be rearranged in a number of different ways to generate an iterative formula. Sometimes different iterative formulae of $f(x) = 0$ may result in different roots of the equation from the same value of x_0.

KEY INFORMATION

Different rearrangements of $f(x) = 0$ giving different iterative formulae may result in different roots.

Example 8

a $x^2 - 7x - 3 = 0$ can be written in the form $x = 7 + \dfrac{3}{x}$.

Show that it can also be written in the form $x = \dfrac{x^2 - 3}{7}$.

b Show that the iterative formulae $x_{n+1} = 7 + \dfrac{3}{x_n}$ and

$x_{n+1} = \dfrac{(x_n)^2 - 3}{7}$ give different roots of the equation to 1 decimal place when $x_0 = 3$.

Solution

a $x^2 - 7x - 3 = 0$

Add $7x$ to both sides of the equation.

$x^2 - 3 = 7x$

Divide both sides of the equation by 7.

$x = \dfrac{x^2 - 3}{7}$

b $x_{n+1} = 7 + \dfrac{3}{x_n}$

$x_0 = 3$

$x_1 = 8$

$x_2 = 7.375$

$x_3 = 7.406\,779\,661$

$x_4 = 7.405\,034\,325$

This gives a root of 7.4 to 1 d.p.

$x_{n+1} = \dfrac{(x_n)^2 - 3}{7}$

$x_0 = 3$

$x_1 = 0.857\,142\,857\,1$

$x_2 = -0.323\,615\,160\,3$

$x_3 = -0.413\,610\,461\,1$

$x_4 = -0.404\,132\,340\,9$

This gives a root of −0.4 to 1 d.p.

Stop and think Is there any relationship between the iterative formula chosen and the root to which it converges?

Sometimes, although a value for x_0 is chosen that is close to a root, repeated iterations may not converge to a root. Instead, they may **diverge**.

KEY INFORMATION

An iterative formula and a value of x_0 close to the root do not always result in approximations converging to the root – instead, they may diverge.

Example 9

The equation $e^x - x - 2 = 0$ has a root in the interval $1 < x < 2$. By using the iterative formula $x_{n+1} = e^{x_n} - 2$ with $x_0 = 2$, find the root to 2 d.p.

Solution

$x_{n+1} = e^{x_n} - 2$

$x_0 = 2$

$x_1 = 5.389\,056\,099$

$x_2 = 216.996\,5769$

$x_3 = 1.739\,465\,981 \times 10^{94}$

This is a divergent sequence, so it is not possible to find the root to 2 d.p. starting with $x_0 = 2$.

Exercise 6.4A

1 $x^2 = 11 - 5x$ can be written in each of the following the forms.

  a $x = \sqrt{11 - 5x}$ **b** $x = \dfrac{11}{x} - 5$ **c** $x = \dfrac{11 - x^2}{5}$

The equation $x^2 + 5x - 11 = 0$ has a root in the interval $1 < x < 2$. Using each of the above forms as the basis for a general iterative formula and with $x_0 = 1$, determine whether each formula results in the same or an alternative root.

2 The equation $x^3 - 6 = x^2$ has a root in the interval $2 < x < 3$. Use the iterative formula

$x_{n+1} = \sqrt{\left(\dfrac{6}{x_n - 1} \right)}$, starting with $x_0 = 5$. Does this combination of iterative formula and starting

value result in a convergent or divergent sequence of results?

3 Show that two different forms of the equation $x^2 + 4x - 7 = 0$ with $x_0 = -5$ result in convergence to two different roots.

4 The equation $2x^3 - 5x + 1 = 0$ has a root in the interval $-2 < x < -1$. Find a value of x_0 such that a divergent sequence of results is produced.

(PS) 5 The equation $x^5 + 3x^2 - 7 = 0$ has a root near $x = 1.5$. Is there a general iterative formula that produces a divergent sequence of results when $x_0 = 1.5$?

(PS) 6 The equation $x^4 - 4x^2 + x = 0$ has a root in the interval $1.8 < x < 1.9$. Find the first integer value of $x_0 > 1.9$ such that a divergent sequence of results is produced.

7 Can the iterative formula $x_{n+1} = \left(x_n^2 + 3x_n - 3 \right)^{\frac{1}{3}}$ be used to find the root of $x^3 - x^2 - 3x + 3 = 0$ in the interval $-2 < x < -1$? You must justify your answer.

8 The equation $x^2 - 3x - 13 = 0$ has a root in the interval $-3 < x < -2$. Determine whether the

iterative formula $x_{n+1} = \dfrac{13}{x_n - 3}$ can be used to find the root in this interval.

9 The equation $x^4 - x^2 - 4 = 0$ as a root in each of the intervals $-2 < x < -1$ and $1 < x < 2$. Using the

iterative formula $x_{n+1} = \left(\dfrac{4}{x_n^2 - 1} \right)^{\frac{1}{2}}$ with $x_0 = -3$, determine to which root, if any, this iterative

formula converges.

10 Determine whether the iterative formula $x_{n+1} = \left(\dfrac{1}{2}(x_n^2 + 8) \right)^{\frac{1}{5}}$ for $2x^5 - x^2 - 8 = 0$, with $x_0 = 0$,

converges to the root.

Mathematics in life and work: Group discussion

You have been asked to write a computer program to use an iterative method to find approximate solutions to the equation $x^3 - 4x - 7 = 0$.

1 How many different iterative formulae could you choose from to use in your computer program? Are any of the formulae better than others? If so, why?

2 How could you determine how many solutions this equation has?

3 What would be a good starting value to use in your chosen iterative formula and why?

SUMMARY OF KEY POINTS

> An iteration is a repeated process or action.

> If there is an interval in which the sign changes, then the interval contains a root of $f(x) = 0$.

> If there are an even number of roots for $f(x) = 0$ in the given interval, then a sign change will not be detected.

> If there is a sign change, check that a root exists by sketching the function and verify that the sign change does not represent a discontinuity.

> x_n is the latest value from the iterative formula and x_{n+1} is the next value.

> To solve an equation in the form $f(x) = 0$ by an iterative method, rearrange $f(x) = 0$ into a form where $x = g(x)$. Then use the formula $x_{n+1} = g(x_n)$ for the iterations.

> If each iteration gets closer to the root, the sequence is convergent.

> Different rearrangements of $f(x) = 0$ giving different iterative formulae may result in different roots.

> An iterative formula and a value of x_0 close to the root do not always result in approximations converging to the root. Instead, the sequence of values may diverge.

EXAM-STYLE QUESTIONS

1 **a** Show that the equation $y = x + \sin x - 3$ has a root in the interval $2.1 < x < 2.3$.

   b Using the iterative formula $x_{n+1} = 3 - \sin(x_n)$ and $x_0 = 2.1$, find x_3 correct to 2 decimal places.

2 Let $f(x) = x^2 + 6x - 13$.

   a Show that $f(x) = 0$ can be written in the form $x = \dfrac{(13 - x^2)}{6}$.

   b Using the iterative formula $x_{n+1} = \dfrac{(13 - x_n^2)}{6}$ and $x_0 = 1$, find x_4 to 3 decimal places.

3 **a** By sketching each of the graphs $y = \dfrac{3}{x}$ and $y = 2^x$ for $-4 \leqslant x \leqslant 4$, show that the equation $\dfrac{3}{x} = 2^x$ has exactly one real root in the interval $1 < x < 2$.

   b Use the iterative formula $x_{n+1} = \dfrac{3}{2^{x_n}}$ and $x_0 = 1$ to find x_3, correct to 3 decimal places.

4 The equation $x^2 - 7x - 3 = 0$ has two reals roots, denoted by α and β, where $\alpha > 0$ and $\beta < 0$.

   a Find, by calculation, the pair of consecutive integers between which each of α and β lies.

   b Use the iterative formula $x_{n+1} = 7 + \dfrac{3}{x_n}$ to determine α correct to 2 decimal places. Give the result of each iteration to 6 decimal places.

PS **5** Use the iterative formula $x_{n+1} = 5 + \dfrac{4}{x_n} - \dfrac{7}{x_n^2}$ to find an approximate solution of
$x^3 - 5x^2 - 4x + 7 = 0$ near to $x = 5$, correct to 2 decimal places.

6 **a** Show that one root of the equation $x^5 + 2x - 7 = 0$ lies in the interval $1 < x < 2$.

    b Show that $x^5 + 2x - 7 = 0$ can be written in the form $x_{n+1} = (a - bx_n)^{\frac{1}{5}}$, where a and b are constants to be found.

    c Find the root of $x^5 + 2x - 7 = 0$ in the interval $1 < x < 2$ to 2 d.p.

    d Show graphically the first two iterations for this formula.

PS **7** The diagram shows the intersection of the curves $y = \cos x$ and $y = x$, where x is in radians. Use the iterative formula $x_{n+1} = \cos x_n$ to find the coordinates of the point of intersection of the two curves, correct to 2 decimal places. Give the result of each iteration to 5 decimal places.

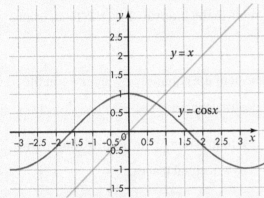

PS **8** The equation $x^3 - x^2 - 6 = 0$ has a root in the interval $2.2 < x < 2.4$. Use the iterative formula $x_{n+1} = \left(x_n^2 + 6\right)^{\frac{1}{3}}$, starting with $x_0 = 2.2$, to find the solution of the equation correct to 3 decimal places. Give the result of each iteration to 6 decimal places.

9 **a** By sketching a graph, show that the equation $x^3 - 3x + 4 = 0$ has exactly one real root, α.

    b Verify by calculation that $-2.5 < \alpha < -2.0$.

    c Use the iterative formula $x_{n+1} = \left(3x_n - 4\right)^{\frac{1}{3}}$ to find α correct to 2 decimal places. Give the result of each iteration to 5 decimal places.

PS **10** The area of the triangle shown is 24 units². Use a general iterative method to find the lengths of the two given sides of the triangle to 2 decimal places.

PS **11** **a** Use the iterative formula $x_{n+1} = \left(x_n^3 + 6\right)^{\frac{1}{4}}$ to find a root of the equation $x^4 - x^3 - 6 = 0$ in the interval $1 < x < 2$, giving your answer correct to 2 decimal places.

    b Show graphically the first two iterations for this formula.

C 12 **a** Use the iterative formula $x_{n+1} = -\sqrt{4 - \dfrac{1}{x_n}}$ with $x_0 = -2$ to find the negative root of the equation $x^4 - 4x^2 + x = 0$, giving your answer correct to 2 decimal places.

  b Show that using the iterative formula $x_{n+1} = \dfrac{4}{x_n} - \dfrac{1}{x_n^2}$ with a suitable starting point does not lead to the same negative root of the equation $x^4 - 4x^2 + x = 0$.

PS 13 Use the iterative formula $x_{n+1} = \left(\dfrac{5x - 1}{2}\right)^{\frac{1}{3}}$ with $x_0 = -1$ to find the negative root of the equation $2x^3 - 5x + 1 = 0$, giving your answer correct to 2 decimal places.

PS 14 Use the iterative formula $x_{n+1} = e^{-x_n^2}$ with $x_0 = -1$ to solve the equation $x^2 + \ln x = 0$ correct to 3 decimal places. You are not required to record the outcome of each iterative step.

15 The equation $x^5 - x^4 + 2x^3 - 3x^2 + \dfrac{1}{2} = 0$ has three roots in the interval $-1 < x < 2$. Find, using algebra, individual intervals for each root.

16 Using the iterative formula $x_{n+1} = (x_n^2 + 1)^{\frac{1}{4}}$, find the root of the equation $x^4 - x^2 - 1 = 0$ in the interval $-2 < x < 2$, giving your answer correct to 2 decimal places.

17 Use the iterative formula $x_{n+1} = \left(\dfrac{5x - 1}{2}\right)^{\frac{1}{3}}$, with $x_0 = 4$, to find one of the positive roots of the equation $2x^3 - 5x + 1 = 0$, giving your answer correct to 3 decimal places.

Mathematics in life and work

A computer programmer is asked to write a program to approximate the solutions to the equation $e^x = 3x + 5$.

1 On the same pair of axes, draw graphs of $y = e^x$ and $y = 3x + 5$ where $-3 \leqslant x \leqslant 3$, highlighting any points of intersection.

2 By choosing two different suitable iterative formulae and starting values, find approximations for each of the solutions to the equation $e^x = 3x + 5$. You must show all your attempts.

3 Comment on the efficiency of the formulae and starting values you choose in **part b**.

7 VECTORS

MATHEMATICS IN LIFE AND WORK

In this chapter, you will learn how **vectors** can be used in two and three dimensions. A vector is a quantity with a size and a direction. Vectors are a key feature of mechanics (such as displacement, velocity, acceleration and forces), but vectors are used in pure mathematics as well. They are used in a variety of careers – for example:

> If you were an aerospace engineer or an astrophysicist, you could model the orbits of satellites, planets, stars and determine the most efficient way of travelling to the Moon or Mars or how to land a buggy on another planet. In fact, vectors are a requirement in any engineering discipline, such as civil, chemical, environmental, electrical or nuclear engineering.

> If you were a football manager, you could use vectors to analyse how your team played during a match, such as tracking the movement (speed and direction) of each player throughout, enabling you to devise tactics for upcoming fixtures.

> If you were a cartographer, you would use vectors to model natural and constructed geographical features to produce a two-dimensional representation of the landscape. Satellite navigation (sat nav) systems could then use this information to program the sat nav to work out the user's current location and, in a matter of seconds, compute the most efficient route – from potentially millions of possibilities – using built-in maps.

In this chapter, you will consider the application of line vectors to satellite navigation.

LEARNING OBJECTIVES

You will learn how to:

> use standard notation for vectors, i.e. $\begin{pmatrix} x \\ y \end{pmatrix}$, $x\mathbf{i} + y\mathbf{j}$, $\begin{pmatrix} x \\ y \\ z \end{pmatrix}$, $x\mathbf{i} + y\mathbf{j} + z\mathbf{k}$, $\overrightarrow{AB}$, $\mathbf{a}$

> carry out addition and subtraction of vectors and multiplication of a vector by a scalar, and interpret these operations in geometrical terms

> find the mid-point of a line as a vector

> calculate the magnitude of a vector, and use unit vectors, displacement vectors and position vectors

> understand the significance of all the symbols used when the equation of a straight line is expressed in the form $\mathbf{r} = \mathbf{a} + t\,\mathbf{b}$, and find the equation of a line, given sufficient information

> determine whether two lines are parallel, intersect or are skew, and find the point of intersection of two lines when it exists

> calculate the scalar product of two vectors and use scalar products in problems involving lines and points

> multiply a vector by a scalar

> determine whether two lines are parallel or perpendicular.

LANGUAGE OF MATHEMATICS

Key words and phrases you will meet in this chapter:

> collinear, column vector, equal vectors, foot, **i j k** notation, intersection, magnitude, mid-point, opposite vectors, origin, parallel vectors, parallelogram law, position vector, relative displacement, scalar product, skew, triangle law, unit vector, vector

PREREQUISITE KNOWLEDGE

You should already know how to:

> apply Pythagoras' theorem and the sine, cosine and tangent ratios for acute angles to the calculation of a side or of an angle of a right-angled triangle

> solve problems using the sine and cosine rules for any triangle and the formula

  area of triangle $= \frac{1}{2}\, ab \sin C$

> solve simple trigonometrical problems in three dimensions including finding the angle between a line and a plane

> describe a translation in terms of a vector

> add and subtract vectors

> multiply a vector by a scalar

> calculate the magnitude of a vector

> use position vectors

> solve simultaneous linear equations in two unknowns

> simplify surds.

You should be able to complete the following questions correctly:

1 Find the distance of the line segment AB given the following pairs of points.

  a $A(6, -10)$ and $B(1, 2)$

  b $A(-5, -4)$ and $B(17, 7)$ (giving the answer in the form $a\sqrt{5}$)

2 Find the size of the angle FHG in the right-angled triangle that has a right angle at G and sides $GH = 8\,\text{cm}$ and $GF = 5\,\text{cm}$. Give the answer correct to 1 d.p.

3 Find the length of the side AB in the triangle ABC that has $AC = 10\,\text{cm}$, $BC = 17\,\text{cm}$ and angle $ACB = 142°$. Give the answer correct to 3 s.f.

4   Given that $\mathbf{a} = \begin{pmatrix} 4 \\ -3 \end{pmatrix}$ and $\mathbf{b} = \begin{pmatrix} -9 \\ 8 \end{pmatrix}$, find:

 a $\mathbf{a} + \mathbf{b}$

 b $3\mathbf{a} - \mathbf{b}$

 c   the magnitude of **a**.

5   Solve the simultaneous equations $5 + 3t = 9 - u$ and $-1 + 4t = 11 + 2u$.

7.1 Definition of a vector

Consider two points $A(5, 6)$ and $B(-19, 16)$ on a metre square grid.

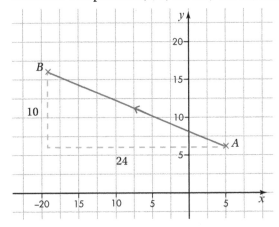

To get from A to B you need to travel 24 m horizontally in the negative x-direction and 10 m vertically in the positive y-direction. The vector $\overrightarrow{AB}$ describes this journey. This is called the displacement vector from A to B or the relative displacement of B from A.

There are two ways of writing the vector $\overrightarrow{AB}$.

The first way of writing the vector is using **i j notation**. **i** is defined as one unit in the positive x-direction (east) and **j** as one unit in the positive y-direction (north). If you have to travel in the negative x- or y-direction, then the coefficient of **i** or **j** will be negative. The vector $\overrightarrow{AB}$ would be written as $(-24\mathbf{i} + 10\mathbf{j})$ m.

> Remember that vectors can also be written using lower-case letters such as **a**, in which case, the vector does not have to have a fixed location.

This can be extended to **i j k notation** when working in three dimensions. For example, the vector $(-\mathbf{i} + 5\mathbf{j} + 2\mathbf{k})$ describes a journey of 1 unit in the negative x-direction, 5 units in the positive y-direction and 2 units in the positive z-direction. If you think of **i** and **j** as east and north on a horizontal plane, then z is the third dimension (up and down).

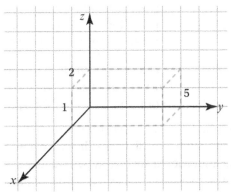

The second way of writing the vector is to use a **column vector**, as you will have seen when translating shapes in your previous study. It is made up of two numbers, one for x and one for y, written vertically in brackets, and they are the same numbers as the coefficients you used to write the vector in **i j** notation.

The vector $\overrightarrow{AB}$ would be written as $\begin{pmatrix} -24 \\ 10 \end{pmatrix}$. Again, this can be

extended to three dimensions, such as $\begin{pmatrix} 18 \\ 5 \\ 2 \end{pmatrix}$.

KEY INFORMATION
Vectors can be written in **i j k** notation or as column vectors.

Adding vectors

To add two vectors in **i j** notation (or written as column vectors), add the **i** terms and the **j** terms separately.

For example, $(3\mathbf{i} + 6\mathbf{j}) + (5\mathbf{i} - 2\mathbf{j}) = (3 + 5)\mathbf{i} + (6 - 2)\mathbf{j} = 8\mathbf{i} + 4\mathbf{j}$.

Using column vectors, this would be written as

$\begin{pmatrix} 3 \\ 6 \end{pmatrix} + \begin{pmatrix} 5 \\ -2 \end{pmatrix} = \begin{pmatrix} 8 \\ 4 \end{pmatrix}$.

Subtraction works in the same way.

For example, $(8\mathbf{i} - 2\mathbf{j}) - (-3\mathbf{i} + 5\mathbf{j}) = (8 - (-3))\mathbf{i} + (-2 - 5)\mathbf{j} = 11\mathbf{i} - 7\mathbf{j}$.

Using column vectors, this would be written as

$\begin{pmatrix} 8 \\ -2 \end{pmatrix} - \begin{pmatrix} -3 \\ 5 \end{pmatrix} = \begin{pmatrix} 11 \\ -7 \end{pmatrix}$.

Similarly, in three dimensions,
$(2\mathbf{i} - 5\mathbf{j} + 8\mathbf{k}) + (3\mathbf{i} + 7\mathbf{j} - 10\mathbf{k}) = 5\mathbf{i} + 2\mathbf{j} - 2\mathbf{k}$ and
$(2\mathbf{i} - 5\mathbf{j} + 8\mathbf{k}) - (3\mathbf{i} + 7\mathbf{j} - 10\mathbf{k}) = -\mathbf{i} - 12\mathbf{j} + 18\mathbf{k}$.

Types of vectors

Equal vectors	Two vectors are equal if they have the same magnitude and the same direction.
Opposite vectors	Two vectors are opposite if they have the same magnitude but point in opposite directions.
Parallel vectors	Two vectors are parallel if they point in the same direction. They do not need to have the same magnitude. One vector will be a multiple of the other or they will share a common factor.

Parallel vectors have a common factor. For example, $(6\mathbf{i} + 8\mathbf{j})$ and $(9\mathbf{i} + 12\mathbf{j})$ can be written as $2(3\mathbf{i} + 4\mathbf{j})$ and $3(3\mathbf{i} + 4\mathbf{j})$, respectively, which have a common factor of $(3\mathbf{i} + 4\mathbf{j})$.

Alternatively, when a vector is written in terms of one or more single letters, such as $\mathbf{a}$, $3\mathbf{b}$ or $\mathbf{a} - 2\mathbf{b}$, with a bold font or by drawing a wavy line underneath such as $\underset{\sim}{a}$, you will use the common factor to prove that two vectors such as $(2\mathbf{a} - 5\mathbf{b})$ and $(4\mathbf{a} - 10\mathbf{b})$ are parallel.

KEY INFORMATION
If two vectors are parallel, then they must contain a common factor.

For example, if you are told that a vector is parallel to $(3\mathbf{i} + 2\mathbf{j})$, it can be written as $k(3\mathbf{i} + 2\mathbf{j})$.

PURE MATHEMATICS 3

Example 1

The vectors **m** and **n** are given by $(5\mathbf{i} - 2\mathbf{j})$ and $(-4\mathbf{i} + \mathbf{j})$, respectively.

If $w\mathbf{m} + 7\mathbf{n}$ is parallel to $(6\mathbf{i} - \mathbf{j})$, find the value of w.

Solution

Substitute **m** and **n**.

$w(5\mathbf{i} - 2\mathbf{j}) + 7(-4\mathbf{i} + \mathbf{j})$

Write in the form $(a\mathbf{i} + b\mathbf{j})$.

$5w\mathbf{i} - 2w\mathbf{j} - 28\mathbf{i} + 7\mathbf{j} = (5w - 28)\mathbf{i} + (-2w + 7)\mathbf{j}$

For this to be parallel to $(6\mathbf{i} - \mathbf{j})$, it must be equal to $k(6\mathbf{i} - \mathbf{j})$ for some value of k.

$(5w - 28)\mathbf{i} + (-2w + 7)\mathbf{j} = k(6\mathbf{i} - \mathbf{j})$

Equate coefficients for **i** and **j**.

i: $5w - 28 = 6k$

j: $-2w + 7 = -k$

Hence $k = 2w - 7$.

Substitute for k in the equation $5w - 28 = 6k$.

$5w - 28 = 6(2w - 7)$

$5w - 28 = 12w - 42$

$\qquad 14 = 7w$

$\qquad w = 2$

Check your answer by substituting $w = 2$ into the expression $(5w - 28)\mathbf{i} + (-2w + 7)\mathbf{j}$.

$(5 \times 2 - 28)\mathbf{i} + (-2 \times 2 + 7)\mathbf{j} = -18\mathbf{i} + 3\mathbf{j} = -3(6\mathbf{i} - \mathbf{j})$, which is parallel to $(6\mathbf{i} - \mathbf{j})$.

Exercise 7.1A

1

a Simplify these vectors.

   i $(4\mathbf{i} + 7\mathbf{j}) + (-2\mathbf{i} + 5\mathbf{j})$

   ii $(-\mathbf{i} + \mathbf{j}) + (8\mathbf{i} - 5\mathbf{j})$

   iii $\begin{pmatrix} 0 \\ -6 \end{pmatrix} + \begin{pmatrix} 13 \\ 2 \end{pmatrix}$

   iv $\begin{pmatrix} -2 \\ 5 \\ 1 \end{pmatrix} + \begin{pmatrix} 7 \\ -8 \\ 5 \end{pmatrix}$

b Simplify these vectors.

   i $(-3\mathbf{i} + 4\mathbf{j}) - (\mathbf{i} + 6\mathbf{j})$

   ii $\begin{pmatrix} 4 \\ 17 \end{pmatrix} - \begin{pmatrix} 4 \\ 7 \end{pmatrix}$

   iii $\begin{pmatrix} -5 \\ -1 \end{pmatrix} - \begin{pmatrix} -3 \\ 10 \end{pmatrix}$

   iv $(7\mathbf{i} - 3\mathbf{j} - 9\mathbf{k}) - (2\mathbf{i} + 7\mathbf{j} - 13\mathbf{k})$

PURE MATHEMATICS 3

c Simplify these vectors.

   i $5(2\mathbf{i} + 4\mathbf{j})$

   ii $\dfrac{1}{2}\begin{pmatrix} 22 \\ 9 \end{pmatrix}$

   iii $-4\begin{pmatrix} 5 \\ -3 \end{pmatrix}$

   iv $8\begin{pmatrix} 9 \\ -3 \\ -5 \end{pmatrix}$

d Simplify these vectors.

   i $(-3\mathbf{i} + 4\mathbf{j}) - 3(-2\mathbf{i} + 5\mathbf{j})$

   ii $3\begin{pmatrix} 7 \\ -4 \end{pmatrix} + 5\begin{pmatrix} -4 \\ 6 \end{pmatrix}$

   iii $\dfrac{3}{4}\begin{pmatrix} -12 \\ 11 \end{pmatrix} - 2\begin{pmatrix} -19 \\ 2 \end{pmatrix}$

C **2** The vectors **a** and **b** are given by $\mathbf{a} = \begin{pmatrix} 4 \\ 1 \end{pmatrix}$ and $\mathbf{b} = \begin{pmatrix} -1 \\ 3 \end{pmatrix}$.

This diagram shows the geometric representation of **a** + **b**.

a **i** Find $\begin{pmatrix} 4 \\ 1 \end{pmatrix} + \begin{pmatrix} -1 \\ 3 \end{pmatrix}$.

   ii Explain how your answer to **part a** is represented in the diagram.

b **i** Find **a** – **b**.

   ii Represent **a** – **b** in a diagram.

   iii Explain the geometric significance of **a** – **b**.

c **i** Find 2**a**.

   ii Represent 2**a** in a diagram.

   iii Explain the geometric significance of 2**a**.

3 Find the values of p and q given that:

a $5\begin{pmatrix} p \\ 2 \end{pmatrix} + q\begin{pmatrix} 8 \\ -10 \end{pmatrix} = \begin{pmatrix} 26 \\ 15 \end{pmatrix}$

b $p\begin{pmatrix} 2 \\ 7 \end{pmatrix} + q\begin{pmatrix} 4 \\ 5 \end{pmatrix} = \begin{pmatrix} 10 \\ 53 \end{pmatrix}$

c $p(4\mathbf{i} - \mathbf{j}) + q(3\mathbf{i} + 5\mathbf{j}) = 18\mathbf{i} + 7\mathbf{j}$.

4 The vectors **a** and **b** are given $\begin{pmatrix} 5 \\ -2 \end{pmatrix}$ and $\begin{pmatrix} 7 \\ 4 \end{pmatrix}$, respectively.

a Find 4**a** – 2**b**.

Given that $2\mathbf{a} - u\mathbf{b} = 2\begin{pmatrix} 19 \\ 6 \end{pmatrix}$,

b find the vector given by $u\mathbf{a} + 3\mathbf{b}$.

C Communication   **MM** Mathematical modelling   **PS** Problem solving

5 Vectors **s** and **t** are given by $\mathbf{s} = \begin{pmatrix} 2 \\ w+47 \\ -4 \end{pmatrix}$ and $\mathbf{t} = \begin{pmatrix} 3 \\ v \\ 2 \end{pmatrix}$.

Given that $k\mathbf{s} - 4\mathbf{t} = \begin{pmatrix} 6 \\ 31 \\ w \end{pmatrix}$, find the value of v.

6 Given that $\mathbf{a} = 4\mathbf{i} - \mathbf{j}$ and $\mathbf{b} = \mathbf{i} + 2\mathbf{j}$, find:

a p if $p\mathbf{a} + 2\mathbf{b}$ is parallel to $(5\mathbf{i} + \mathbf{j})$

b q if $3\mathbf{a} + q\mathbf{b}$ is parallel to $(\mathbf{i} - \mathbf{j})$.

(PS) 7 Find values for a and b such that $a\begin{pmatrix} 2 \\ -1 \end{pmatrix} + b\begin{pmatrix} 8 \\ -3 \end{pmatrix}$:

a is equal to $\begin{pmatrix} 6 \\ -4 \end{pmatrix}$

b is parallel to the x-axis and has a magnitude of 12

c is parallel to $y = -x$.

8 Vectors **m** and **n** are given by $\begin{pmatrix} a+3 \\ a-1 \end{pmatrix}$ and $\begin{pmatrix} 3 \\ a \end{pmatrix}$, respectively.

Given that $a\mathbf{m} - 5\mathbf{n} = 3\begin{pmatrix} b \\ -3 \end{pmatrix}$, find the value of b.

(PS) 9 The vectors **a** and **b** are given by $\begin{pmatrix} 3 \\ -1 \end{pmatrix}$ and $\begin{pmatrix} 1 \\ 3 \end{pmatrix}$, respectively.

Given that $p\mathbf{a} + q\mathbf{b}$ is parallel to $\begin{pmatrix} 5 \\ 3 \end{pmatrix}$, show that $7p = 6q$.

(PS) 10 The vector $\mathbf{c} = \begin{pmatrix} 2 \\ n \end{pmatrix}$ and the vector $\mathbf{d} = \begin{pmatrix} 4 \\ 13 \end{pmatrix}$.

Given that $8\mathbf{c} + 3\mathbf{d} = \begin{pmatrix} k \\ 79 \end{pmatrix}$,

a find the value of k

b find the value of n.

Given that $4\mathbf{c} + h\mathbf{d}$ is parallel to the line $y = 2x$,

c find the value of h.

Stop and think
Why does **Question 7 part c** have an infinite number of answers but **parts a** and **b** only have one each?

Can you find values for a and b such that the vector is parallel to other straight lines that pass through the origin such as $y = x$, $y = 2x$ and the y-axis?

7.2 Vector geometry

A vector can be labelled in various ways. If a vector connects two points such as A and B then it is often written as $\overrightarrow{AB}$. The arrow shows that the vector is the journey from A to B, whereas $\overrightarrow{BA}$ would be the journey from B to A.

Alternatively, a vector can be written in terms of one or more single letters such as $\mathbf{a}$, $3\mathbf{b}$ or $\mathbf{a} - 2\mathbf{b}$. In this case, the vector is not fixed between two specific points, and in a parallelogram $OABC$, where $\overrightarrow{OA} = \mathbf{a}$, the vector $\overrightarrow{CB}$ will also be equal to $\mathbf{a}$, as will any other vector that is parallel and the same length.

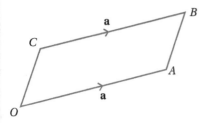

Triangle and parallelogram laws

Vectors can be added using the **triangle law**. If the vector $\overrightarrow{AB}$ ($\mathbf{a}$) is followed by the vector $\overrightarrow{BC}$ ($\mathbf{b}$), then the result is the vector $\overrightarrow{AC}$ ($\mathbf{a} + \mathbf{b}$).

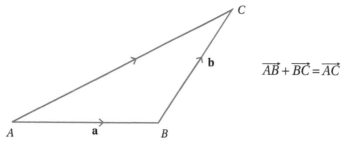

$$\overrightarrow{AB} + \overrightarrow{BC} = \overrightarrow{AC}$$

By drawing the two vectors in the opposite order on the same diagram ($\mathbf{b}$ followed by $\mathbf{a}$), you complete a parallelogram. This is called the **parallelogram law**. $\overrightarrow{AC}$ is the diagonal of the parallelogram. The parallelogram is useful because it highlights the fact that the order of adding vectors is *not* important, as $\mathbf{a} + \mathbf{b}$ gives the same result as $\mathbf{b} + \mathbf{a}$.

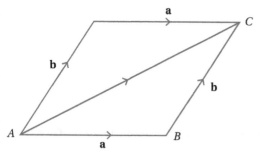

Example 2

OMN is a triangle. $\overrightarrow{OM} = 14\mathbf{m}$; $\overrightarrow{ON} = 21\mathbf{n}$.

P lies on *OM* such that $OP : PM = 2 : 5$.

Q is a point such that $\overrightarrow{ON} = \frac{7}{2}\overrightarrow{OQ}$.

Prove that *PQ* is parallel to *MN*.

Solution

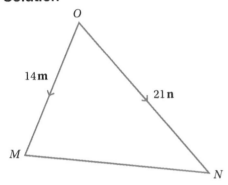

To prove that two vectors are parallel, you need to show that they have a common factor.

In **Example 2,** this means finding both vectors in terms of **m** and **n** and then factorising each one to show what the common factor is.

In order to prove that the vectors $\overrightarrow{PQ}$ and $\overrightarrow{MN}$ are parallel, you need to show that they have a common factor.

$$\overrightarrow{MN} = \overrightarrow{MO} + \overrightarrow{ON} = -14\mathbf{m} + 21\mathbf{n}$$
$$= 7(-2\mathbf{m} + 3\mathbf{n}).$$

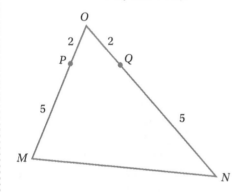

Annotate the diagram.

$$\overrightarrow{OP} = \frac{2}{7}\overrightarrow{OM} = \frac{2}{7} \times 14\mathbf{m} = 4\mathbf{m}.$$

Since $\overrightarrow{ON} = \frac{7}{2}\overrightarrow{OQ}$, $\overrightarrow{OQ} = \frac{2}{7}\overrightarrow{ON} = \frac{2}{7} \times 21\mathbf{n} = 6\mathbf{n}.$

$$\overrightarrow{PQ} = \overrightarrow{PO} + \overrightarrow{OQ} = -4\mathbf{m} + 6\mathbf{n}.$$
$$= 2(-2\mathbf{m} + 3\mathbf{n}).$$

Since $\overrightarrow{MN}$ and $\overrightarrow{PQ}$ have a common factor, they are parallel.

PURE MATHEMATICS 3

The mid-point rule

Consider the triangle OAB with $\overrightarrow{OA} = \mathbf{a}$ and $\overrightarrow{OB} = \mathbf{b}$. C is the **mid-point** of AB.

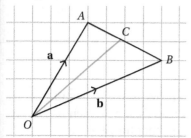

How can you find the vector $\overrightarrow{OC}$ in terms of $\mathbf{a}$ and $\mathbf{b}$?

Start with $\overrightarrow{AB} = \overrightarrow{AO} + \overrightarrow{OB} = -\mathbf{a} + \mathbf{b}$.

$\overrightarrow{AC} = \frac{1}{2}\overrightarrow{AB}$, so $\overrightarrow{AC} = \frac{1}{2}(-\mathbf{a} + \mathbf{b})$.

$\overrightarrow{OC} = \overrightarrow{OA} + \overrightarrow{AC} = \mathbf{a} + \frac{1}{2}(-\mathbf{a} + \mathbf{b}) = \mathbf{a} - \frac{1}{2}\mathbf{a} + \frac{1}{2}\mathbf{b} = \frac{1}{2}\mathbf{a} + \frac{1}{2}\mathbf{b} = \frac{1}{2}(\mathbf{a} + \mathbf{b})$.

> An alternative method would have been to use $\overrightarrow{OC} = \overrightarrow{OB} + \overrightarrow{BC}$

Hence the mid-point of AB is given by $\frac{1}{2}(\overrightarrow{OA} + \overrightarrow{OB})$.

Three points on the same straight line are described as **collinear**.

Example 3

$UVWX$ is a parallelogram with $\overrightarrow{UV} = 3\mathbf{a}$ and $\overrightarrow{UX} = 8\mathbf{b}$.

Y is the point between V and W such that VY is three times as long as YW.

Z is the point such that $\overrightarrow{XZ} = \frac{4}{3}\overrightarrow{XW}$.

Prove that U, Y and Z are collinear.

> To prove three points are collinear, it is sufficient to show that two vectors joining any of the three points are parallel, since any two of these vectors must contain a common point.

Solution

Start by drawing a diagram.

Since $UVWX$ is a parallelogram, $\overrightarrow{VW} = \overrightarrow{UX} = 8\mathbf{b}$.

Y is on the line segment VW and splits the line in the ratio $3 : 1$.

Hence $\overrightarrow{VY} = \frac{3}{4}\overrightarrow{VW} = \frac{3}{4} \times 8\mathbf{b} = 6\mathbf{b}$.

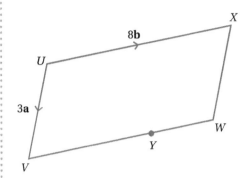

You need to find any two vectors from $\overrightarrow{UY}$, $\overrightarrow{UZ}$ and $\overrightarrow{YZ}$.

$\overrightarrow{UY} = \overrightarrow{UV} + \overrightarrow{VY} = 3\mathbf{a} + 6\mathbf{b} = 3(\mathbf{a} + 2\mathbf{b})$.

Add point Z to the diagram. Z lies on the line segment XW extended such that $\overrightarrow{XZ} = \frac{4}{3}\overrightarrow{XW}$. Since U, Y and Z are collinear, it will also lie on the line UY when that is extended.

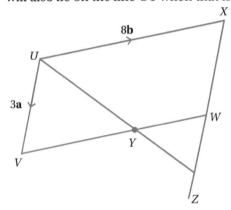

Since $UVWX$ is a parallelogram, $\overrightarrow{XW} = \overrightarrow{UV} = 3\mathbf{a}$.

$\overrightarrow{UZ} = \overrightarrow{UX} + \overrightarrow{XZ} = 8\mathbf{b} + \frac{4}{3} \times 3\mathbf{a} = 8\mathbf{b} + 4\mathbf{a} = 4(\mathbf{a} + 2\mathbf{b})$.

Since $\overrightarrow{UY}$ and $\overrightarrow{UZ}$ have a common factor, they are parallel.

Since $\overrightarrow{UY}$ and $\overrightarrow{UZ}$ also have a common point, U, Y and Z are collinear.

You can also find $\overrightarrow{YZ}$.

$\overrightarrow{YZ} = \overrightarrow{YU} + \overrightarrow{UX} + \overrightarrow{XZ}$

$= -3\mathbf{a} - 6\mathbf{b} + 8\mathbf{b} + 4\mathbf{a}$

$= \mathbf{a} + 2\mathbf{b}$

Hence $\overrightarrow{YZ}$ is also parallel to $\overrightarrow{UY}$ and $\overrightarrow{UZ}$.

PURE MATHEMATICS 3

Exercise 7.2A

1 The triangle RST is such that $\overrightarrow{RS} = 4\mathbf{a}$ and $\overrightarrow{RT} = 6\mathbf{b}$.

U is the mid-point of RS and V is the mid-point of RT.

Prove that $\overrightarrow{ST}$ is parallel to $\overrightarrow{UV}$.

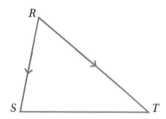

C **2** $OABC$ is a parallelogram. $\overrightarrow{OA} = 6\mathbf{a}$ and $\overrightarrow{OC} = 6\mathbf{c}$.

S is the centre of $OABC$.

T is the mid-point of the side BC.

The point U divides AB in the ratio $2 : 1$.

a Find each vector in terms of $\mathbf{a}$ and $\mathbf{c}$.

   i $\overrightarrow{OB}$ **ii** $\overrightarrow{AC}$ **iii** $\overrightarrow{OU}$

   iv $\overrightarrow{TA}$ **v** $\overrightarrow{OS}$ **vi** $\overrightarrow{US}$

   vii $\overrightarrow{UT}$ **viii** $\overrightarrow{ST}$

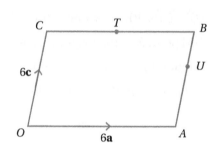

b What is the relationship between the vectors $\overrightarrow{ST}$ and $\overrightarrow{AB}$?

3 DEF is a triangle with $\overrightarrow{DE} = 12\mathbf{a}$ and $\overrightarrow{DF} = 8\mathbf{b}$.

M divides DE in the ratio $1 : 2$ and N divides DF in the ratio $1 : 2$.

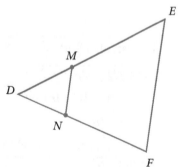

a Prove that $FEMN$ is a trapezium.

b Given that DEF has an area of 72 units^2, find the area of $FEMN$.

4 In the parallelogram $OABC$, $\overrightarrow{OA} = \mathbf{a}$ and $\overrightarrow{OC} = \mathbf{c}$.

D is the mid-point of the side OA. E is the mid-point of the side AB.

a Prove that $\overrightarrow{DE}$ is parallel to $\overrightarrow{OB}$.

The side OA is extended to F such that $OA = AF$.

b Prove that C, E and F are collinear.

5 Refer to the diagram in **Question 2**, reproduced below.

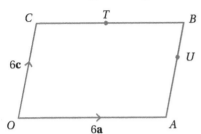

Given that X is the point such that $\overrightarrow{OA} = \overrightarrow{AX}$, prove that the points T, U and X are collinear.

6 In the diagram, $ABCD$ is a quadrilateral such that $\overrightarrow{AB} = 15\mathbf{a}$, $\overrightarrow{DC} = 5\mathbf{a}$ and $\overrightarrow{DA} = 14\mathbf{b}$.
AD is extended to the point E such that the line segments AD and DE are in the ratio $2 : 1$.
Prove that the points B, C and E are collinear.

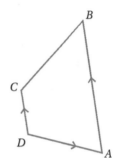

7 In triangle OAB, $OM : MA = ON : NB = 2 : 1$.
$\overrightarrow{OA} = \mathbf{a}$ and $\overrightarrow{OB} = \mathbf{b}$.
X is the intersection of MB and NA.

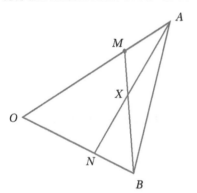

a Show that $\overrightarrow{OX} = \dfrac{2}{5}(\mathbf{a} + \mathbf{b})$.

It is now given that $OM : MA = ON : NB = p : q$.

b Prove that $\overrightarrow{OX} = \dfrac{p}{2p+q}(\mathbf{a} + \mathbf{b})$.

8 In the diagram, $\overrightarrow{JK} = 4\mathbf{a}$, $\overrightarrow{LM} = 6\mathbf{a}$ and $\overrightarrow{JM} = 15\mathbf{b}$.

a Explain why *JKN* and *MLN* are similar triangles.

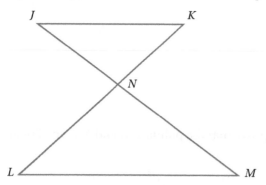

b The point *P* lies on the line *MN*, or on *MJ* produced. Given that the line segment *MP* is twice as long as the line segment *PN*, find the two possible vectors $\overrightarrow{LP}$.

9 *DEFG* is a parallelogram such that $\overrightarrow{DE} = 12\mathbf{a}$ and $\overrightarrow{DG} = 8\mathbf{b}$.

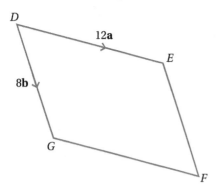

J is the mid-point of *DG*.

K is a point on *FG* such that $\overrightarrow{FG} = 3\overrightarrow{KG}$.

Given that *JKQ* and *EFQ* are both straight lines, find the vector $\overrightarrow{DQ}$ in terms of **a** and **b**.

10 *OPQR* is a parallelogram. *T* divides the side *RQ* in the ratio 5 : 1.

S is the centre of the parallelogram.

$\overrightarrow{OP} = 6\mathbf{p}$.

$\overrightarrow{OR} = 4\mathbf{r}$.

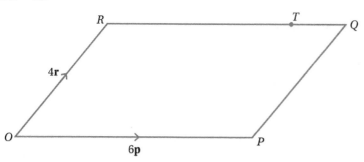

Given that *PQU* and *STU* are straight lines, find the vector $\overrightarrow{OU}$ in terms of **p** and **r**.

7.3 Magnitude of a vector

The **magnitude** (or length) of a vector can be found using Pythagoras' theorem. The vector is the hypotenuse of the right-angled triangle for which the **i** and **j** (and **k**) coefficients are the shorter sides. The magnitude of the vector $\overrightarrow{AB}$ is often written as $\left|\overrightarrow{AB}\right|$, with the vector between two vertical lines. Consider the

vector $\overrightarrow{AB} = \begin{pmatrix} -24 \\ 10 \end{pmatrix}$ from **Section 7.1**.

$$c^2 = a^2 + b^2$$

$$c = \sqrt{a^2 + b^2}$$

$$= \sqrt{(-24)^2 + 10^2}$$

$$= \sqrt{576 + 100}$$

$$= \sqrt{676}$$

$$= 26\,m$$

The magnitude $\left|\overrightarrow{AB}\right| = 26\,m$.

> **KEY INFORMATION**
>
> Use Pythagoras' theorem to calculate the magnitude (length) of a vector.

You can also use the formula for finding the distance between two points.

$$d^2 = (x_1 - x_2)^2 + (y_1 - y_2)^2.$$

Let $A(5, 6) = (x_1, y_1)$ and $B(-19, 16) = (x_2, y_2)$.

$$d^2 = (5 - (-19))^2 + (6 - 16)^2$$

$$= (24)^2 + (-10)^2$$

$$= 576 + 100$$

$$= 676$$

$$d = \sqrt{676}$$

$$= 26\,m$$

This formula can also be used to find the distance between two points in three dimensions.

$$d^2 = (x_1 - x_2)^2 + (y_1 - y_2)^2 + (z_1 - z_2)^2$$

PURE MATHEMATICS 3

Example 4

Find the distance between the points (11, –2, 56) and
(–41, 70, 17).

Solution

$d^2 = (11 - (-41))^2 + (-2 - 70)^2 + (56 - 17)^2$

$\quad = (52)^2 + (-72)^2 + (39)^2$

$\quad = 2704 + 5184 + 1521$

$\quad = 9409$

$d = \sqrt{9409}$

$\quad = 97$

In two dimensions, the direction of the vector can be described
as an angle. This is often measured from the positive **i** vector.
For vectors with a positive **j** component, the angle is measured
anticlockwise from **i** and is positive ($0° < \theta < 180°$). For vectors with
a negative **j** component, the angle is measured clockwise from **i**
and is negative ($-180° < \theta < 0°$).

KEY INFORMATION

The positive **i** vector has
the same direction as the
positive x-axis.

Start by finding the acute angle inside the right-angled triangle.

Angle = $\tan^{-1}\left(\dfrac{10}{24}\right) = 22.6°$.

The angle you need is the obtuse angle measured anticlockwise
from **i**.

$180° - 22.6° = 157.4°$

KEY INFORMATION

Use trigonometry to find the
direction of a vector.

A **unit vector** is a vector with a magnitude of 1. The vector $\overrightarrow{AB}$ has a
magnitude of 26 m but you can find the unit vector in the direction
of ($-24\mathbf{i} + 10\mathbf{j}$) by dividing the vector by 26. Hence the unit vector is
$\dfrac{1}{26}(-24\mathbf{i} + 10\mathbf{j})$ m.

KEY INFORMATION

A unit vector has a
magnitude of 1.

You can use the unit vector to find vectors of other magnitudes
in the same direction. For example, if you wished to find a vector
with a magnitude of 182 m in the direction of ($-24\mathbf{i} + 10\mathbf{j}$), you
could multiply the unit vector by 182.

$\dfrac{1}{26}(-24\mathbf{i} + 10\mathbf{j}) \times 182 = 7(-24\mathbf{i} + 10\mathbf{j})$ m

Mathematics in life and work: Group discussion

You are modelling roads for a satellite navigation system. A particular road junction has three roads meeting at one point. Road A joins the junction with a vector of $(2\mathbf{i} + 5\mathbf{j})$ m. Roads B and C leave the junction with vectors $(-6\mathbf{i} + \mathbf{j})$ m and $(4\mathbf{i} + 2\mathbf{j})$ m, respectively.

1 Find the distance travelled from the beginning of road A to the end of road B.

2 Find the distance travelled if you entered the junction from road B and left via road C.

3 How could this model be improved?

4 How would you use this type of information to devise the most efficient route between two towns?

5 How could you use this type of information to calculate an estimated time of arrival for a long journey?

Example 5

For the vector $(12\mathbf{i} - 8\mathbf{j})$, find:

a the magnitude of the vector

b the angle the vector makes with $\mathbf{i}$.

Drawing a rough diagram may help you visualise the situation in the question.

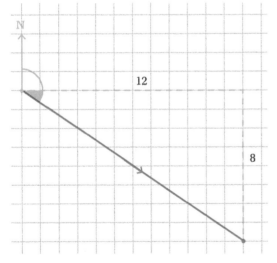

Solution

a Use Pythagoras' theorem.

$$c = \sqrt{a^2 + b^2}$$

$$= \sqrt{12^2 + (-8)^2}$$

$$= \sqrt{208}$$

$$= 4\sqrt{13} \text{ units.}$$

The magnitude of the vector is $4\sqrt{13}$ units.

b Use trigonometry.

Angle $= \tan^{-1}\left(\dfrac{8}{12}\right) = 33.7°$.

The vector makes an angle of $-33.7°$ with the vector **i**.

> Notice the angle is negative as it is below the positive **i** axis.

Example 6

Find a vector parallel to $(8\mathbf{i} - 12\mathbf{j} - 9\mathbf{k})$ with magnitude 68.

Solution

Find the unit vector in the direction $8\mathbf{i} - 12\mathbf{j} - 9\mathbf{k}$.

$$\text{Magnitude} = \sqrt{8^2 + (-12)^2 + (-9)^2}$$

$$= \sqrt{289}$$

$$= 17.$$

Hence unit vector $= \dfrac{1}{17}(8\mathbf{i} - 12\mathbf{j} - 9\mathbf{k})$.

Finally, multiply the unit vector by 68.

$$\dfrac{1}{17}(8\mathbf{i} - 12\mathbf{j} - 9\mathbf{k}) \times 68 = 4(8\mathbf{i} - 12\mathbf{j} - 9\mathbf{k})$$

$$= 32\mathbf{i} - 48\mathbf{j} - 36\mathbf{k}$$

> Alternatively, once you know the magnitude is 17, you can compare it with the magnitude of 68.
>
> Since $68 \div 17 = 4$, the vector you need is $4(8\mathbf{i} - 12\mathbf{j} - 9\mathbf{k})$.

Exercise 7.3A

1 Find the angle between each vector and the vector **i**, giving all answers correct to 1 decimal place.

    a $7\mathbf{i} + 4\mathbf{j}$ **b** $-10\mathbf{i} + 7\mathbf{j}$ **c** $\begin{pmatrix} -6 \\ -5 \end{pmatrix}$ **d** $\begin{pmatrix} 4 \\ -9 \end{pmatrix}$

2 **a** Find the magnitude of the vector, giving the answer, where appropriate, in the form $a\sqrt{b}$, where b is prime.

    i $\begin{pmatrix} 182 \\ -120 \end{pmatrix}$ **ii** $12\mathbf{i} - 8\mathbf{j} - 9\mathbf{k}$ **iii** $46\mathbf{i} - 46\mathbf{j} + 23\mathbf{k}$

    iv $\begin{pmatrix} 5 \\ -4 \\ 3 \end{pmatrix}$ **v** $\begin{pmatrix} -7 \\ -7 \\ -7 \end{pmatrix}$

b Find the unit vector in each direction.

   i $-4\mathbf{i} + 3\mathbf{j}$ **ii** $143\mathbf{i} - 24\mathbf{j}$ **iii** $-72\mathbf{i} + 33\mathbf{j} + 56\mathbf{k}$

   iv $\begin{pmatrix} 11 \\ 13 \\ 8 \end{pmatrix}$ **v** $\begin{pmatrix} 5 \\ -4 \\ -2 \end{pmatrix}$

(C) 3 Rebekah and Henry are discussing a vector of the form $\begin{pmatrix} a \\ a \end{pmatrix}$.

Rebekah says that the vector has a bearing of 045°. Henry says that the vector has a bearing of 225°.

   a Who is correct? Explain your answer.

   b What form would a vector have to take to have a bearing of 315°?

(PS) 4 The vector $\begin{pmatrix} a \\ b \end{pmatrix}$ has a magnitude of $\sqrt{130}$.

If a and b are integers, how many possible pairs of values are there for a and b?

5 In each case find a vector of the given magnitude and in the given direction.

   a magnitude = 52, direction = $-5\mathbf{i} + 12\mathbf{j}$

   b magnitude = 195, direction = $63\mathbf{i} + 16\mathbf{j}$

   c magnitude = 12, direction = $\mathbf{i} - 2\mathbf{j} + 2\mathbf{k}$

   d magnitude = 14, direction = $3\mathbf{i} - 6\mathbf{j} + 2\mathbf{k}$

   e magnitude = $9\sqrt{26}$, direction = $3\mathbf{i} - \mathbf{j} + 4\mathbf{k}$

(MM) (C) 6 A hotel is set out as a coordinate grid in a cube measuring 12 units along each edge. The manager of the hotel has insisted that no bedroom is permitted to be further than 9 units from a bathroom. The lifts along one vertical edge are modelled as the origin for each floor. Room 168 is on Floor 7 and is modelled as $(2\mathbf{i} + 9\mathbf{j})$. The two closest bathrooms to Room 168 are on Floor 5 at $(7\mathbf{i} + \mathbf{j})$ and Floor 10 at $(8\mathbf{i} + 3\mathbf{j})$.

   a Is the manager's rule satisfied for Room 168? Explain your answer.

   b State an assumption that have you made in answering this question.

(MM) 7 A ship sails 14 km east, then 9 km southeast.

   a Calculate the displacement of the ship from its starting point.

   b Write the displacement of the ship in the form $(a\mathbf{i} + b\mathbf{j})$ km, where a and b are exact values.

   c Calculate the bearing of the ship from its starting point.

8 Vectors **a** and **b** are given by $(-3\mathbf{i} + 4\mathbf{j} + 12\mathbf{k})$ and $(7\mathbf{i} + 39\mathbf{j} - 2\mathbf{k})$, respectively.

   a Find the distance of **a** from the origin.

   b Find the distance between **a** and **b**.

   c Given that $3\mathbf{a} + c\mathbf{b} = -23\mathbf{i} - 66\mathbf{j} + 40\mathbf{k}$, find the value of c.

9 The point $A(7, -3, 3)$ lies on the surface of the sphere S_1, which has its centre at the origin.

   a Does point $B(2, -8, -1)$ lie inside, outside or on the surface of S_1?

   Points $J(-4, 6, -1)$ and $K(-2, 2, 17)$ are such that JK is a diameter of the sphere S_2.

   b Verify that the points $L(2, 10, 3)$ and $M(4, -2, 7)$ lie on the sphere.

10 The magnitude of the vector $\begin{pmatrix} u-3 \\ u+1 \\ u-2 \end{pmatrix}$ is 17.

   Find the value of u.

> **Stop and think**
>
> The cosine rule could be used to find the distance in **Question 8**. What other ways are there of finding the distance?

7.4 Position vectors

In **Section 7.1**, you were shown that the relative displacement of B from A describes how to get from the point A to the point B. This section will show how to use **position vectors** to find the same result, then extend it to solve problems in pure mathematics.

The position vector of a point is its position relative to the **origin**. For $A(5, 6)$, the position vector is $(5\mathbf{i} + 6\mathbf{j})$, and for $B(-19, 16)$, the position vector is simply $(-19\mathbf{i} + 16\mathbf{j})$.

The relative displacement of B from A, or $\overrightarrow{AB}$, can be found by subtracting the position vector of A from the position vector of B in the same way that you subtracted vectors in **Section 7.2**.

$(-19\mathbf{i} + 16\mathbf{j}) - (5\mathbf{i} + 6\mathbf{j}) = -19\mathbf{i} - 5\mathbf{i} + 16\mathbf{j} - 6\mathbf{j} = -24\mathbf{i} + 10\mathbf{j}$

Hence you can find the distance between A and B by finding the magnitude of the relative displacement using Pythagoras' theorem, which, as you have seen, is 26.

> **KEY INFORMATION**
>
> To find the relative displacement of B from A, $\overrightarrow{AB}$, subtract the position vector of A from the position vector B.

How would a satellite navigation system make use of position vectors and displacement vectors when deciding which route is the most efficient?

Example 7

Point A has coordinates $(-3, 14)$ and point B has position vector $\begin{pmatrix} u \\ -1 \end{pmatrix}$.

a Write A as a position vector.

b Given that the magnitude of the vector $\overrightarrow{AB}$ is 17, find both possible values of u.

Solution

a The position vector of A is $\begin{pmatrix} -3 \\ 14 \end{pmatrix}$.

b Find the vector $\overrightarrow{AB}$.

$$\overrightarrow{AB} = \mathbf{b} - \mathbf{a} = \begin{pmatrix} u \\ -1 \end{pmatrix} - \begin{pmatrix} -3 \\ 14 \end{pmatrix} = \begin{pmatrix} u+3 \\ -15 \end{pmatrix}$$

The magnitude of $\overrightarrow{AB}$ is given by

$$\sqrt{(u+3)^2 + (-15)^2} = \sqrt{u^2 + 6u + 9 + 225} = \sqrt{u^2 + 6u + 234}.$$

$$\sqrt{u^2 + 6u + 234} = 17$$

$$u^2 + 6u + 234 = 17^2 = 289$$

$$u^2 + 6u - 55 = 0$$

$$(u + 11)(u - 5) = 0$$

$$u = -11 \text{ or } 5.$$

> Use Pythagoras' theorem to find the magnitude of the vector.

Why are there two possible position vectors for B?

If both answers for u were plotted as points B and C, what could you say about triangle ABC?

Example 8

The points L, M and N have position vectors $\mathbf{l} = (4\mathbf{i} - 10\mathbf{j} + 3\mathbf{k})$, $\mathbf{m} = (-3\mathbf{i} - 6\mathbf{j} + \mathbf{k})$ and $\mathbf{n} = (5\mathbf{i} + 3\mathbf{j} - 9\mathbf{k})$, respectively.

a Prove that LMN is a right-angled triangle.

b Find the area of the triangle LMN, giving the answer in the form $a\sqrt{345}$.

Solution

a $\overrightarrow{LM} = \mathbf{m} - \mathbf{l} = (-3\mathbf{i} - 6\mathbf{j} + \mathbf{k}) - (4\mathbf{i} - 10\mathbf{j} + 3\mathbf{k}) = (-7\mathbf{i} + 4\mathbf{j} - 2\mathbf{k})$

Magnitude $= \sqrt{(-7)^2 + 4^2 + (-2)^2} = \sqrt{69}$.

$\overrightarrow{LN} = \mathbf{n} - \mathbf{l} = (5\mathbf{i} + 3\mathbf{j} - 9\mathbf{k}) - (4\mathbf{i} - 10\mathbf{j} + 3\mathbf{k}) = (\mathbf{i} + 13\mathbf{j} - 12\mathbf{k})$

Magnitude $= \sqrt{1^2 + 13^2 + (-12)^2} = \sqrt{314}$.

$\overrightarrow{MN} = \mathbf{n} - \mathbf{m} = (5\mathbf{i} + 3\mathbf{j} - 9\mathbf{k}) - (-3\mathbf{i} - 6\mathbf{j} + \mathbf{k}) = (8\mathbf{i} + 9\mathbf{j} - 10\mathbf{k})$

Magnitude $= \sqrt{8^2 + 9^2 + (-10)^2} = \sqrt{245}$.

Since $LM^2 + MN^2 = LN^2$ ($69 + 245 = 314$), Pythagoras'
theorem is satisfied and LMN is a right-angled triangle.

b $A = \frac{1}{2}bh$

$= \frac{1}{2} \times \sqrt{69} \times \sqrt{245}$

$= \frac{1}{2} \times \sqrt{69} \times 7\sqrt{5}$

$= \frac{7}{2} \times \sqrt{345}$

> To show that *LMN* is a right-angled triangle, it is sufficient to demonstrate that the three sides satisfy Pythagoras' theorem.

> In two dimensions, you could also show that the triangle is right-angled using the gradients of the shorter sides, which will be perpendicular and hence have a product of -1.

Exercise 7.4A

1 The position vectors **a**, **b** and **c**, given by $\begin{pmatrix} 3 \\ 8 \\ 2 \end{pmatrix}$, $\begin{pmatrix} 5 \\ -1 \\ 3 \end{pmatrix}$ and $\begin{pmatrix} -4 \\ 3 \\ -9 \end{pmatrix}$ respectively, represent the points A, B and C.

Find the relative displacement of:

a A from B

b C from A

c B from A

d B from C

2 Four buildings have position vectors as follows:

hospital: $(-\mathbf{i} + 7\mathbf{j})$ km

supermarket: $(5\mathbf{i} + 2\mathbf{j})$ km

town hall: $(3\mathbf{i} - 9\mathbf{j})$ km

museum: $(-2\mathbf{i} - 4\mathbf{j})$ km.

a For which two buildings is the relative displacement:

i $(2\mathbf{i} + 11\mathbf{j})$ km

ii $(7\mathbf{i} + 6\mathbf{j})$ km

iii $(4\mathbf{i} - 16\mathbf{j})$ km?

b Find the relative displacement of the hospital from the museum.

c Which two buildings are $5\sqrt{2}$ km apart?

3 Points D, E and F are given by the position vectors $(11\mathbf{i} + 5\mathbf{j} - 4\mathbf{k})$, $(7\mathbf{i} - 3\mathbf{j} + 4\mathbf{k})$ and $(15\mathbf{i} + \mathbf{j} + 12\mathbf{k})$, respectively.

Prove that DEF is a right angle.

4 For points A, B with position vectors $\mathbf{a} = \begin{pmatrix} 20 \\ 23 \\ -15 \end{pmatrix}$ and $\mathbf{b} = \begin{pmatrix} -8 \\ 100 \\ 29 \end{pmatrix}$,

  a find the mid-point of AB

  b find the distance between A and B.

5 Points A, B, C and D are given by the position vectors $\begin{pmatrix} 6 \\ 7 \\ -2 \end{pmatrix}$, $\begin{pmatrix} -11 \\ 8 \\ 7 \end{pmatrix}$, $\begin{pmatrix} 9 \\ -2 \\ 12 \end{pmatrix}$ and $\begin{pmatrix} 18 \\ 1 \\ 1 \end{pmatrix}$, respectively.

  a Find the vector $\overrightarrow{AD}$.

  b Find the vector $\overrightarrow{BC}$.

  c Hence deduce what type of quadrilateral $ABCD$ is.

C 6 $D(2, -8, 5)$, $E(-11, 3, -7)$ and $F(-6, -1, 2)$ are three vertices of a triangle. Prove that the triangle DEF is isosceles.

PS 7 The position vector of S is $(q\mathbf{i} - 6\mathbf{j})$ m.

The position vector of T is $(-2\mathbf{i} + 3\mathbf{j})$ m.

  a If $q = 4$, calculate the distance between S and T, giving the answer in the form $a\sqrt{13}$ m.

  b Given that $ST = 15$ m, find both possible values of q.

C 8 The points T, U and V have position vectors $(10\mathbf{i} + 8\mathbf{j} - 3\mathbf{k})$, $(13\mathbf{i} + 12\mathbf{j} - 14\mathbf{k})$ and $(3\mathbf{i} + 27\mathbf{j} + 2\mathbf{k})$, respectively.

Prove that the triangle TUV is:

  a scalene   **b** right-angled.

PS 9 G and H have position vectors $\begin{pmatrix} -2 \\ c+1 \\ -8 \end{pmatrix}$ and $\begin{pmatrix} c-1 \\ -4 \\ c+2 \end{pmatrix}$, respectively, such that the magnitude of GH is 19.

  a Show that $3c^2 + 32c - 235 = 0$.

  b Given that $c > 0$, find the distance of H from the origin.

PS 10 Point P has position vector $(43\mathbf{i} + 145\mathbf{j} + 383\mathbf{k})$ and point Q has position vector $(-65\mathbf{i} - 11\mathbf{j} + 32\mathbf{k})$.

  a Find the length of the line PQ.

Point R has position vector $(169\mathbf{i} + 61\mathbf{j} - 72\mathbf{k})$.

  b Show that PQR is a right angle.

  c Find the area of triangle PQR.

PURE MATHEMATICS 3

7.5 Vector equation of a straight line

You should already know the general Cartesian equation of a straight line in two dimensions, $y = mx + c$. m represents the gradient, which is another way of describing the direction of the line. c represents a point on the line, the y-intercept, but in theory any point on the line could be used.

Consider the line L which passes through the points A and X. The point A (given by position vector $\overrightarrow{OA}$) is a point on the line and the vector $\overrightarrow{AX}$ is the direction of the line.

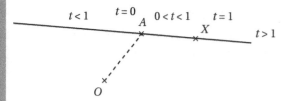

Hence the vector equation of a line can be written in the form $\mathbf{r} = \mathbf{a} + t\mathbf{b}$, where $\mathbf{a}$ is a known point on the line, such as $\mathbf{A}$, and $\mathbf{b}$ is the direction of the line, such as $\overrightarrow{AX}$. t is a variable. For different values of t, you can travel to different points on the line. When $t = 0$, the point is A. When $t = 1$, the point is X. For other values of t the diagram shows you where you would travel to.

$\mathbf{r}$ is the general position vector of any point on the line for different values of t.

Given two points on the line, you can find $\mathbf{b}$ by subtracting one position vector from the other. You can also remove any common factors.

If you are going to use two (or more) straight line equations at the same time, you need to use a different variable for each line. Use t for the first line and u for the second. For example, if your first line uses t, such as $(3\mathbf{i} - \mathbf{j} + 10\mathbf{k}) + t(2\mathbf{j} - 7\mathbf{k})$, then your second line should use u instead, such as $(-5\mathbf{i} - 2\mathbf{j} + 19\mathbf{k}) + u(\mathbf{i} - 3\mathbf{j} + \mathbf{k})$.

> **KEY INFORMATION**
>
> The vector equation of a line can be written in the form $\mathbf{r} = \mathbf{a} + t\mathbf{b}$, where $\mathbf{a}$ is a known point on the line and $\mathbf{b}$ is the direction of the line.

> t and u have been used here to represent the variables in the vector equation of a straight line. Other pairs of letters are sometimes used instead. For example λ and μ.

Example 9

a Find the vector equation of the line l_1 through $U(5, -2, 18)$ and $V(19, 5, -10)$, writing your answer using column vectors.

b Find the point on the line l_1 which has $z = 30$.

Solution

a The direction of the line l_1 is given by $\overrightarrow{UV}$.
$\overrightarrow{UV} = (19\mathbf{i} + 5\mathbf{j} - 10\mathbf{k}) - (5\mathbf{i} - 2\mathbf{j} + 18\mathbf{k}) = 14\mathbf{i} + 7\mathbf{j} - 28\mathbf{k}$
This can be simplified, by dividing through by 7, to $2\mathbf{i} + \mathbf{j} - 4\mathbf{k}$.

> Note that direction vectors can be simplified in this way because it is the direction, not the magnitude, that you need. You cannot, however, simplify a position vector.

PURE MATHEMATICS 3

Hence the line l_1 can be written as
$\mathbf{r} = (5\mathbf{i} - 2\mathbf{j} + 18\mathbf{k}) + t(2\mathbf{i} + \mathbf{j} - 4\mathbf{k})$.

b l_1 can be rewritten as $\mathbf{r} = (5 + 2t)\mathbf{i} + (-2 + t)\mathbf{j} + (18 - 4t)\mathbf{k}$.

When $z = 30$, $18 - 4t = 30$.

$$4t = -12$$

$$t = -3$$

When $t = -3$, $\mathbf{r} = (5 + 2(-3))\mathbf{i} + (-2 + (-3))\mathbf{j} + 30\mathbf{k}$.

$\mathbf{r} = -\mathbf{i} - 5\mathbf{j} + 30\mathbf{k}$

> When you have three coordinates, they are written as (x, y, z), and hence $z = 18 - 4t$ in this question.

Stop and think   What other points could be used instead of U for the point on the line?

Exercise 7.5A

1   Find a vector equation of the line that passes through the point with position vector **a** and direction **b**.

a $\mathbf{a} = 5\mathbf{i} + 3\mathbf{j}$, $\mathbf{b} = -7\mathbf{i} + \mathbf{j}$

b $\mathbf{a} = -2\mathbf{i} + 8\mathbf{k}$, $\mathbf{b} = -3\mathbf{i} + 2\mathbf{j} - \mathbf{k}$

c $\mathbf{a} = \begin{pmatrix} 9 \\ -2 \end{pmatrix}$, $\mathbf{b} = \begin{pmatrix} 4 \\ 3 \end{pmatrix}$

d $\mathbf{a} = \begin{pmatrix} 10 \\ 3 \\ -6 \end{pmatrix}$, $\mathbf{b} = \begin{pmatrix} 0 \\ -1 \\ 3 \end{pmatrix}$

2   **a**   Find a vector equation of the line l_1 through $G(3, 7, -9)$ and $H(2, -4, 3)$.

b   Find a vector equation of the line l_2 through $P(-30, -6, 30)$ and $Q(-26, -12, 33)$.

3   Find a vector equation of the line l that passes through $A\begin{pmatrix} 11 \\ -6 \\ 9 \end{pmatrix}$ and $B\begin{pmatrix} 1 \\ 4 \\ 29 \end{pmatrix}$. Your direction vector should be simplified.

4   Line l is given by the equation $\mathbf{r} = \begin{pmatrix} 11 \\ 5 \end{pmatrix} + t\begin{pmatrix} 7 \\ -2 \end{pmatrix}$.

a   Show that the point with position vector $\begin{pmatrix} 39 \\ -3 \end{pmatrix}$ lies on the line l.

b   Find the position vector for the point on the line l with $t = -3$.

c   Given that the point $(v, -3v)$ lies on l, find the value of v.

5 The line l passes through the points with position vectors $\begin{pmatrix} -2 \\ 9 \\ 1 \end{pmatrix}$ and $\begin{pmatrix} 10 \\ 5 \\ -7 \end{pmatrix}$.

a Find an equation for the line l in the form $\mathbf{r} = \mathbf{a} + t\mathbf{b}$ (where $\mathbf{b}$ is simplified).

b Show that the point $\begin{pmatrix} 19 \\ 2 \\ -13 \end{pmatrix}$ lies on the line l.

c Find the missing numbers (a to g) for these points that lie on the line l.

   i $\mathbf{i} + a\mathbf{j} - \mathbf{k}$

   ii $-8\mathbf{i} + b\mathbf{j} + c\mathbf{k}$

   iii $d\mathbf{i} + e\mathbf{j} - 3\mathbf{k}$

   iv $f\mathbf{i} + 4\mathbf{j} + g\mathbf{k}$

d Find the point on the line l which also lies in the x–z plane.

6 $A(13, -10, 15)$ and $B(22, -22, 9)$ lie on line l with equation $\mathbf{r} = \begin{pmatrix} 7 \\ -2 \\ 19 \end{pmatrix} + t\begin{pmatrix} 3 \\ -4 \\ -2 \end{pmatrix}$.

Point C also lies on l such that it is twice as far from A as from B.

Find all possible coordinates of the point C.

7 Points U and V have respective position vectors $(p\mathbf{i} - 3\mathbf{j} - \mathbf{k})$ and $(2\mathbf{i} + 5\mathbf{j} - 3\mathbf{k})$.

a Find a vector equation for the line l that passes through U and V.

Given that the point $(7, 1, q)$ lies on the line l, find:

b the value of p

c the value of q.

8 The position vectors $\mathbf{a} = 2\mathbf{i} + p\mathbf{j} + q\mathbf{k}$, $\mathbf{b} = -4\mathbf{i} + 13\mathbf{j} - \mathbf{k}$ and $\mathbf{c} = 14\mathbf{i} + \mathbf{j} + 5\mathbf{k}$ represent the points A, B and C. A, B and C are collinear.

a Find the values of p and q.

b Find the coordinates of the point where the line through A, B and C intersects the y–z plane.

c Find the distance AC, writing the answer in the form $a\sqrt{14}$.

PS **9** Find the points on the line $\mathbf{r} = \begin{pmatrix} 3 \\ -2 \\ -6 \end{pmatrix} + t\begin{pmatrix} 1 \\ 2 \\ 3 \end{pmatrix}$ that are 29 units from the origin.

10 The line l has direction vector $(3\mathbf{i} + 6\mathbf{j} - 2\mathbf{k})$. The point A with position vector $\mathbf{a} = -10\mathbf{i} + 4\mathbf{j} + \mathbf{k}$ lies on l.

a Show that the point B with position vector $\mathbf{b} = 5\mathbf{i} + 34\mathbf{j} - 9\mathbf{k}$ also lies on l.

b Find the magnitude of the vector $\overrightarrow{AB}$.

The point C has position vector $\mathbf{c} = 8\mathbf{i} - 2\mathbf{j} + 10\mathbf{k}$.

c Show that ABC is a right-angled triangle.

d Find the area of the triangle ABC.

7.6 Parallel, intersecting and skew lines

In two dimensions, lines must either **intersect** or be **parallel**.

In three dimensions, lines can intersect or be parallel to each other, but they can also be **skew**, which means that they do not intersect even though they are not parallel.

Parallel vectors

As described in **Section 7.1**, two vectors which are parallel are multiples of each other.

So, for example, $6\mathbf{i} + 8\mathbf{j} - 4\mathbf{k}$ and $9\mathbf{i} + 12\mathbf{j} - 6\mathbf{k}$ are parallel because $6\mathbf{i} + 8\mathbf{j} - 4\mathbf{k} = 2(3\mathbf{i} + 4\mathbf{j} - 2\mathbf{k})$ and $9\mathbf{i} + 12\mathbf{j} - 6\mathbf{k} = 3(3\mathbf{i} + 4\mathbf{j} - 2\mathbf{k})$.

Assuming that the directions have already been factorised, you can tell if two lines are parallel if their direction vectors are the same.

Intersecting and skew lines

In two dimensions, two non-parallel lines will always intersect.

In three dimensions this is not the case.

For two lines to intersect, there needs to be a value of t and a value of u for which all of $\mathbf{i}$, $\mathbf{j}$ and $\mathbf{k}$ are the same.

Consider two lines, l_1 and l_2.

l_1 is given by $\mathbf{r} = \mathbf{a} + t\mathbf{b}$ and l_2 is given by $\mathbf{r} = \mathbf{c} + u\mathbf{d}$.

For the two lines to intersect, there must be values of t and u such that $\mathbf{a} + t\mathbf{b} = \mathbf{c} + u\mathbf{d}$.

Note that there will be three equations (one each for $\mathbf{i}$, $\mathbf{j}$ and $\mathbf{k}$). To find the values of t and u, put the coefficients from two of the equations equal to each other and solve the equations simultaneously. After you have found t and u, you need to ensure that the values of t and u you have found also work in the third equation. If the values of t and u do not satisfy the third equation, then the lines are skew.

> **KEY INFORMATION**
>
> In two dimensions, lines are either parallel or they intersect.
>
> In three dimensions, lines can be parallel, intersect or be skew to each other.

> Recall that the vector equation of a line includes a variable such as t or u.

> **KEY INFORMATION**
>
> If there are no values for t and u that satisfy all three equations in $\mathbf{i}$, $\mathbf{j}$ and $\mathbf{k}$, then the lines are skew.

PURE MATHEMATICS 3

PURE MATHEMATICS 3

Example 10

The line l_1 passes through the points $A(3\mathbf{i} - 3\mathbf{j} - 6\mathbf{k})$ and $B(9\mathbf{i} - 15\mathbf{j} + 3\mathbf{k})$.

The line l_2 has the equation $\mathbf{r} = \begin{pmatrix} 4 \\ 7 \\ 0 \end{pmatrix} + t \begin{pmatrix} 1 \\ 2 \\ 3 \end{pmatrix}$.

Find the point of intersection of the lines l_1 and l_2.

Solution

Write the equation of each line in the form $\mathbf{r} = a\mathbf{i} + b\mathbf{j} + c\mathbf{k}$.

The direction of $l_1 = \overrightarrow{AB} = (9\mathbf{i} - 15\mathbf{j} + 3\mathbf{k}) - (3\mathbf{i} - 3\mathbf{j} - 6\mathbf{k})$
$= 6\mathbf{i} - 12\mathbf{j} + 9\mathbf{k}$.

This can be simplified to $2\mathbf{i} - 4\mathbf{j} + 3\mathbf{k}$.

$6\mathbf{i} - 12\mathbf{j} + 9\mathbf{k} = 3(2\mathbf{i} - 4\mathbf{j} + 3\mathbf{k})$

The equation of l_1 is given by $(3\mathbf{i} - 3\mathbf{j} - 6\mathbf{k}) + u(2\mathbf{i} - 4\mathbf{j} + 3\mathbf{k})$
$= (3 + 2u)\mathbf{i} + (-3 - 4u)\mathbf{j} + (-6 + 3u)\mathbf{k}$.

The equation of l_2 is given by $(4\mathbf{i} + 7\mathbf{j}) + t(\mathbf{i} + 2\mathbf{j} + 3\mathbf{k})$
$= (4 + t)\mathbf{i} + (7 + 2t)\mathbf{j} + (3t)\mathbf{k}$.

When you have more than one straight line equation, use a different variable.

Equate coefficients of $\mathbf{i}$: $3 + 2u = 4 + t$ ①

Equate coefficients of $\mathbf{j}$: $-3 - 4u = 7 + 2t$ ②

Equate coefficients of $\mathbf{k}$: $-6 + 3u = 3t$ ③

① × 2:

$6 + 4u = 8 + 2t$

$-3 - 4u = 7 + 2t$ ②

$9 + 8u = 1$

$8u = -8$

$u = -1$

This is a set of three simultaneous equations with two unknowns. You can solve any pair but you need to check that the values of t and u satisfy the third equation as well.

Substitute $u = -1$ into equation (1) to find t.

$3 + 2(-1) = 4 + t$

$t = -3$

Check the values of t and u in equation ③.

$-6 + 3(-1) = 3(-3)$

Since both sides equal the same value (-9), the values of t and u satisfy all three equations.

To confirm that the point you have found is a point of intersection, check the third equation also holds for the values of t and u found.

Using l_1, the position vector of the point of intersection is given by $(3 + 2(-1))\mathbf{i} + (-3 - 4(-1))\mathbf{j} + (-6 + 3(-1))\mathbf{k} = \mathbf{i} + \mathbf{j} - 9\mathbf{k}$

Therefore, the point of intersection is $(1, 1, -9)$.

What conclusion could you have drawn if the values of t (-3) and u (-1) had not satisfied the third equation in **Example 10**?

Example 11

Show that the lines $(14\mathbf{i} + 5\mathbf{j} - 6\mathbf{k}) + t(4\mathbf{i} + 3\mathbf{j} - 5\mathbf{k})$ and $(3\mathbf{i} + 11\mathbf{j} + 2\mathbf{k}) + u(\mathbf{i} - 4\mathbf{j} + 3\mathbf{k})$ are skew.

Solution

Equate coefficients of **i**: $14 + 4t = 3 + u$ ①

Equate coefficients of **j**: $5 + 3t = 11 - 4u$ ②

Equate coefficients of **k**: $-6 - 5t = 2 + 3u$ ③

① × 4:

$$56 + 16t = 12 + 4u$$

Again, you can use any pair of equations.

$$5 + 3t = 11 - 4u \qquad ②$$

$$61 + 19t = 23$$

$$19t = -38$$

$$t = -2$$

Substitute $t = -2$ into equation ① to find u.

$$14 + 4(-2) = 3 + u$$

$$u = 3$$

Check the values of t and u in equation ③.

Left-hand side $= -6 - 5(-2) = -6 + 10 = 4$.

Right-hand side $= 2 + 3(3) = 11$.

Since the two sides do not equal the same value, the lines are skew.

Exercise 7.6A

1 Find the value of a that makes l_1 and l_2 parallel.

l_1 $\mathbf{r} = 3\mathbf{i} + 2\mathbf{j} + \mathbf{k} + t(4\mathbf{i} + a\mathbf{j} - 12\mathbf{k})$

l_2 $\mathbf{r} = -2\mathbf{i} + 4\mathbf{j} + 3\mathbf{k} + u(3\mathbf{i} - 2\mathbf{j} - 9\mathbf{k})$

C 2 Identify which of these four pairs of vectors are parallel, justifying your choice.

a $\begin{pmatrix} 15 \\ 10 \\ -20 \end{pmatrix}$ and $\begin{pmatrix} 8 \\ 12 \\ 24 \end{pmatrix}$
   b $\begin{pmatrix} 27 \\ -42 \\ -36 \end{pmatrix}$ and $\begin{pmatrix} 9 \\ -16 \\ -12 \end{pmatrix}$

c $\begin{pmatrix} -6 \\ -4 \\ 10 \end{pmatrix}$ and $\begin{pmatrix} -3 \\ -1 \\ 13 \end{pmatrix}$
   d $\begin{pmatrix} 12 \\ -8 \\ 32 \end{pmatrix}$ and $\begin{pmatrix} 21 \\ -14 \\ 56 \end{pmatrix}$

3 Find the point of intersection of the lines l_1 and l_2 from **Exercise 7.5A Question 2**.

4 Find the point of intersection of the lines $\mathbf{r} = \begin{pmatrix} -2 \\ 2 \\ -1 \end{pmatrix} + t \begin{pmatrix} 1 \\ 2 \\ -1 \end{pmatrix}$ and $\mathbf{r} = \begin{pmatrix} 4 \\ 4 \\ -7 \end{pmatrix} + u \begin{pmatrix} -1 \\ 3 \\ 1 \end{pmatrix}$.

5 For each part of this question, line l_1 passes through E and F and line l_2 passes through G and H. Determine whether each pair of lines is parallel, intersecting or skew. If the lines intersect, find the position vector of the point of intersection.

   a $E(17, 6, 34)$, $F(5, 9, 16)$, $G(1, 21, -2)$, $H(-13, -14, 19)$

   b $E(-3, -5, 16)$, $F(12, 0, 1)$, $G(7, 2, 11)$, $H(1, -22, 23)$

   c $E(3, -5, -7)$, $F(-11, 9, 14)$, $G(8, -1, -8)$, $H(2, 5, 1)$

6 Points $A(-8\mathbf{i} + 6\mathbf{j} + \mathbf{k})$ and $B(-2\mathbf{i} + 10\mathbf{j} - 3\mathbf{k})$ lie on the straight line l_1.

   a Find a vector equation for the line l_1.

   Points $C(-3\mathbf{i} - 2\mathbf{j} + 3\mathbf{k})$ and $D(-4\mathbf{i} + 3\mathbf{j} + \mathbf{k})$ lie on the straight line l_2.

   b Write the vector equation for the line l_2 in the form $\mathbf{r} = \mathbf{a} + u\mathbf{b}$.

   c Find the point of intersection, E, of the lines l_1 and l_2.

   d Find the ratio of $AE : EB$.

7 Points $P(-5\mathbf{i} + 2\mathbf{j} + 4\mathbf{k})$ and $Q(-2\mathbf{i} - 3\mathbf{j} + 3\mathbf{k})$ lie on the straight line l_1.

   a Find a vector equation for the line l_1.

   Points $R(-8\mathbf{j} - \mathbf{k})$ and $S(12\mathbf{i} - 23\mathbf{j} + 8\mathbf{k})$ lie on the straight line l_2.

   b Show that l_1 and l_2 are skew lines.

   Point T has position vector $(12\mathbf{i} - 23\mathbf{j} + a\mathbf{k})$. Points R and T lie on the straight line l_3.

   Given that l_1 and l_3 intersect, find:

   c the value of a

   d the position vector of the point at which the lines intersect.

8 Points $A(-3\mathbf{i} - \mathbf{j} + 12\mathbf{k})$ and $B(-2\mathbf{i} + 11\mathbf{k})$ lie on the line l_1.

   The line l_2 is parallel to l_1 and passes through the point C with position vector $(11\mathbf{i} + 9\mathbf{j} + 12\mathbf{k})$.

   a Find the equation of l_2.

   C and $D(15\mathbf{i} + 11\mathbf{j} + 15\mathbf{k})$ lie on the line l_3.

   b Find the point of intersection of the lines l_1 and l_3.

9 Line l_1 passes through the points Q and S with position vectors $\mathbf{q} = \begin{pmatrix} -6 \\ 14 \\ -19 \end{pmatrix}$ and $\mathbf{s} = \begin{pmatrix} -4 \\ 6 \\ -3 \end{pmatrix}$.

Line l_2 is parallel to the line with vector equation $\mathbf{r} = \begin{pmatrix} 0 \\ -6 \\ 1 \end{pmatrix} + t \begin{pmatrix} 4 \\ 7 \\ 4 \end{pmatrix}$ and passes through the

point $T(1, 9, 9)$.

U is the point of intersection of the lines l_1 and l_2.

Show that the triangle STU is isosceles.

 10 Points C, D, E and F have respective position vectors $\begin{pmatrix} -11 \\ 9 \\ 15 \end{pmatrix}$, $\begin{pmatrix} 2 \\ -5 \\ 17 \end{pmatrix}$, $\begin{pmatrix} -1 \\ 4 \\ 5 \end{pmatrix}$ and $\begin{pmatrix} -5 \\ 9 \\ 3 \end{pmatrix}$.

    a Find a vector equation for the line l_1 that passes through the points C and E.

    b Find a vector equation for the line l_2 that passes through the points D and F.

    c Find the position vector of the point G, where l_1 and l_2 intersect.

    d Show that CGF is a right angle.

    e Find the area of the quadrilateral $CDEF$.

7.7 Scalar product

To find the angle (θ) between two vectors (**a** and **b**), you can use the **scalar product**:

$\mathbf{a} \cdot \mathbf{b} = |\mathbf{a}||\mathbf{b}| \cos \theta$

In this formula, **a** and **b** are the directions of the two vectors.

To find $\mathbf{a} \cdot \mathbf{b}$, multiply the **i** coefficients, the **j** coefficients and the **k** coefficients and add the results.

The magnitudes of **a** and **b** are found as before.

If $\cos \theta$ is positive, then the angle is acute. If $\cos \theta$ is negative, then the angle is obtuse, but you can subtract the angle from 180° to find the acute angle. As you can see from the graph below, $\cos \theta = -\cos (180 - \theta)$.

> **KEY INFORMATION**
>
> Use the scalar product, $\mathbf{a} \cdot \mathbf{b} = |\mathbf{a}||\mathbf{b}| \cos \theta$, to find the angle between two vectors.

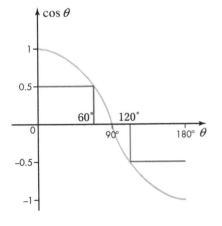

Example 12

l_1 is given by $\mathbf{r} = \begin{pmatrix} 6 \\ -3 \\ 4 \end{pmatrix} + t \begin{pmatrix} 8 \\ 2 \\ -3 \end{pmatrix}$ and l_2 is given by $\mathbf{r} = \begin{pmatrix} 3 \\ 3 \\ 5 \end{pmatrix} + u \begin{pmatrix} -4 \\ 9 \\ -1 \end{pmatrix}$.

Find the acute angle between lines l_1 and l_2, giving the answer to 1 d.p.

Solution

$\mathbf{a}$ and $\mathbf{b}$ are the directions of the lines l_1 and l_2.

Hence $\mathbf{a} = \begin{bmatrix} 8 \\ 2 \\ -3 \end{bmatrix}$ and $\mathbf{b} = \begin{bmatrix} -4 \\ 9 \\ -1 \end{bmatrix}$.

$\mathbf{a} \bullet \mathbf{b} = 8 \times -4 + 2 \times 9 + -3 \times -1 = -32 + 18 + 3 = -11$

The magnitude of $\mathbf{a} = \sqrt{8^2 + 2^2 + (-3)^2} = \sqrt{64 + 4 + 9} = \sqrt{77}$.

The magnitude of $\mathbf{b} = \sqrt{(-4)^2 + 9^2 + (-1)^2} = \sqrt{16 + 81 + 1} = \sqrt{98}$.

$\cos\theta = \dfrac{\mathbf{a} \bullet \mathbf{b}}{|\mathbf{a}||\mathbf{b}|} = \dfrac{-11}{\sqrt{77}\sqrt{98}} = -\dfrac{\sqrt{154}}{98}$

> If $\mathbf{a} \bullet \mathbf{b}$ is negative, then the angle you find will be obtuse. If you need to find the acute angle, subtract the obtuse angle from 180°.

$\theta = \cos^{-1}\left(-\dfrac{\sqrt{154}}{98}\right) = 97.3°$

$\theta = 180° - 97.3° = 82.7°$, correct to 1 d.p.

Example 13

E, F and G have position vectors $\mathbf{e} = 11\mathbf{i} + 6\mathbf{j}$, $\mathbf{f} = 15\mathbf{i} + 4\mathbf{j}$ and $\mathbf{g} = 18\mathbf{i} - 8\mathbf{j}$.

Find the exact area of the triangle EFG.

Solution

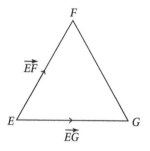

The area of a triangle is given by $A = \frac{1}{2}ab\sin C$.

a and b are the magnitudes of $\mathbf{a}$ and $\mathbf{b}$.

C is the angle between $\mathbf{a}$ and $\mathbf{b}$.

Use the sides EF and EG and the angle FEG. $\mathbf{a} = \overrightarrow{EF}$ and $\mathbf{b} = \overrightarrow{EG}$.

$$\overrightarrow{EF} = (15\mathbf{i} + 4\mathbf{j}) - (11\mathbf{i} + 6\mathbf{j}) = 4\mathbf{i} - 2\mathbf{j}$$

$$\overrightarrow{EG} = (18\mathbf{i} - 8\mathbf{j}) - (11\mathbf{i} + 6\mathbf{j}) = 7\mathbf{i} - 14\mathbf{j}$$

$$\overrightarrow{EF} \bullet \overrightarrow{EG} = 4 \times 7 + -2 \times -14 = 28 + 28 = 56$$

The magnitude of $\overrightarrow{EF} = \sqrt{4^2 + (-2)^2} = \sqrt{16 + 4} = \sqrt{20} = 2\sqrt{5}$.

The magnitude of $\overrightarrow{EG} = \sqrt{7^2 + (-14)^2} = \sqrt{49 + 196} = \sqrt{245} = 7\sqrt{5}$.

$$\cos FEG = \frac{\mathbf{a} \bullet \mathbf{b}}{|\mathbf{a}||\mathbf{b}|} = \frac{56}{2\sqrt{5} \times 7\sqrt{5}} = \frac{4}{5}$$

$$\sin FEG = \sqrt{1 - \left(\frac{4}{5}\right)^2} = \frac{3}{5}$$

Note that it is not necessary to find the angle itself: you can use the trigonometric identity $\sin^2 x + \cos^2 x \equiv 1$. This will be a more accurate method than if you round the angle.

Area $= \frac{1}{2}ab\sin C = \frac{1}{2} \times 2\sqrt{5} \times 7\sqrt{5} \times \frac{3}{5}$.

Area $= 21$ units^2.

Perpendicular lines

To check if two vectors are perpendicular, you can use the scalar product.

If $\theta = 90°$, then $\cos\theta = 0$ and therefore $\mathbf{a} \bullet \mathbf{b} = 0$.

Again, to check if two lines are perpendicular, you need to use the direction vectors of the lines.

KEY INFORMATION

If $\mathbf{a} \bullet \mathbf{b} = 0$, then the vectors are perpendicular.

Example 14

Show that the lines $\mathbf{r} = 4\mathbf{i} - \mathbf{j} - \mathbf{k} + t(3\mathbf{i} - 2\mathbf{j} + 8\mathbf{k})$ and $\mathbf{r} = 2\mathbf{i} - 3\mathbf{j} + 2\mathbf{k} + u(-2\mathbf{i} + 5\mathbf{j} + 2\mathbf{k})$ are perpendicular.

Solution

$\mathbf{a} \bullet \mathbf{b} = (3\mathbf{i} - 2\mathbf{j} + 8\mathbf{k}) \bullet (-2\mathbf{i} + 5\mathbf{j} + 2\mathbf{k})$

$= 3 \times -2 + -2 \times 5 + 8 \times 2$

$= -6 - 10 + 16 = 0$

Since $\mathbf{a} \bullet \mathbf{b} = 0$, $\cos\theta = 0$ and $\theta = 90°$.

Hence the lines are perpendicular.

Exercise 7.7A

1 Find the acute angle between $(4\mathbf{i} - 2\mathbf{j} + 5\mathbf{k})$ and $(3\mathbf{i} + 4\mathbf{j} - 2\mathbf{k})$.

2 Line l_1 has the equation $\mathbf{r} = (3\mathbf{j} - 8\mathbf{k}) + t(7\mathbf{i} - 19\mathbf{j} + 2\mathbf{k})$.
Line l_2 has the equation $\mathbf{r} = (\mathbf{i} - 8\mathbf{j} - 2\mathbf{k}) + u(5\mathbf{i} + 3\mathbf{j} + 11\mathbf{k})$.
Show that l_1 and l_2 are perpendicular.

3 The line l_1 passes through the points $(-3\mathbf{i} + 5\mathbf{j} + 7\mathbf{k})$ and $(2\mathbf{i} - \mathbf{j} + 4\mathbf{k})$.
The line l_2 passes through the points $(\mathbf{i} + 2\mathbf{j} - \mathbf{k})$ and $(6\mathbf{i} - 5\mathbf{j} + 3\mathbf{k})$.
Find the acute angle between the lines.

4 Find the cosine of the angle OAB for which the position
vectors of A and B are $\mathbf{a} = 3\mathbf{i} + 5\mathbf{j} - 4\mathbf{k}$ and $\mathbf{b} = 7\mathbf{i} + 4\mathbf{j} - 3\mathbf{k}$,
respectively.

> If you were asked for the
> cosine of the angle OAB, you
> would use the vectors
> $\overrightarrow{AO}$ and $\overrightarrow{AB}$.

5 Find the exact area of the triangle ABC where $\overrightarrow{AB} = 4\mathbf{i} + 9\mathbf{j} + \mathbf{k}$ and $\overrightarrow{AC} = 9\mathbf{i} + \mathbf{j} + 4\mathbf{k}$.

6 **a** Given that $\mathbf{a} = \mathbf{i} + 3u\mathbf{j} - 2\mathbf{k}$ and $\mathbf{b} = -u\mathbf{i} + u\mathbf{j} + 2\mathbf{k}$, and that $\mathbf{a} \bullet \mathbf{b} = 10$, find both possible
values for u.

b Given also that $u < 0$, show that the cosine of the angle between $\mathbf{a}$ and $\mathbf{b}$ can be written as $\dfrac{5}{\sqrt{123}}$.

7 **a** Points A, B and C are given by $A(2, -3)$, $B(10, -7)$ and $C(14, -15)$.

  i Use the scalar product to find the cosine of the angle ABC.

  ii Hence use the formula $A = \dfrac{1}{2}ab\sin C$ to show that the area of the triangle ABC is 24 unit^2.

b Points D, E and G are given by $D(4, -2)$, $E(1, 2)$ and $G(6, 7)$.
  Use the scalar product to find the area of parallelogram $DEFG$.

8 Points A, B and D have position vectors $\begin{pmatrix} -4 \\ 13 \\ 22 \end{pmatrix}$, $\begin{pmatrix} 4 \\ 17 \\ 14 \end{pmatrix}$ and $\begin{pmatrix} 2 \\ -29 \\ 7 \end{pmatrix}$, respectively.

a Prove that AB and AD are perpendicular.

b Find the position vector of C such that $ABCD$ is a rectangle.

c Find the area of the rectangle $ABCD$.

d Find the position vector of the centre of the rectangle.

(PS) 9 Find a vector perpendicular to both $\begin{pmatrix} 1 \\ 4 \\ -2 \end{pmatrix}$ and $\begin{pmatrix} 2 \\ -4 \\ -7 \end{pmatrix}$.

10 The points A, B, C and D have the following position vectors.

$A\ (15\mathbf{i} - 2\mathbf{j} + \mathbf{k})$

$B\ (11\mathbf{i} - 2\mathbf{j} + 3\mathbf{k})$

$C\ (9\mathbf{i} + 13\mathbf{j} + 19\mathbf{k})$

$D\ (-\mathbf{i} - 12\mathbf{j} - \mathbf{k})$.

a Find the equation of the line l_1 that passes through A and B.

b Find the equation of the line l_2 that passes through C and D.

c Find the point of intersection P of lines l_1 and l_2.

d Find the angle between the lines l_1 and l_2.

e Find the lengths of AP and CP in the form $a\sqrt{5}$.

f Find the area of the triangle APC.

11 Line l_1 has equation $\mathbf{r} = \begin{pmatrix} -8 \\ 16 \\ 1 \end{pmatrix} + t \begin{pmatrix} 7 \\ -3 \\ -6 \end{pmatrix}$ and line l_2 has equation $\mathbf{r} = \begin{pmatrix} -10 \\ 31 \\ -26 \end{pmatrix} + u \begin{pmatrix} 3 \\ -6 \\ 7 \end{pmatrix}$.

The two lines intersect at point P.

Point Q lies on l_1 with $t = 3$ and point R lies on l_2 such that $u > 3$. $PQ = PR$.

a Show that $\cos RPQ = -\dfrac{3}{94}$.

b Show that the triangle PQR has an area of $2\sqrt{8827}$.

Example 15

A pyramid $ABCDE$ has a square base $ABCD$. Its vertices are given by $A(1, 1, -1)$, $B(9, -1, -3)$, $C(9, -7, 3)$, $D(1, -5, 5)$ and $E(8, 3, 7)$. The centre of $ABCD$ is M.

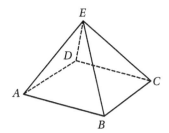

a Find the size of $\angle AEC$, in degrees, correct to 1 decimal place.

b Show that $\angle AME = 90°$.

c Find the volume of $ABCDE$.

Solution

a Use the scalar product.

$$\vec{EA} = (\mathbf{i} + \mathbf{j} - \mathbf{k}) - (8\mathbf{i} + 3\mathbf{j} + 7\mathbf{k}) = -7\mathbf{i} - 2\mathbf{j} - 8\mathbf{k}$$

$$\vec{EC} = (9\mathbf{i} - 7\mathbf{j} + 3\mathbf{k}) - (8\mathbf{i} + 3\mathbf{j} + 7\mathbf{k}) = \mathbf{i} - 10\mathbf{j} - 4\mathbf{k}$$

$$\vec{EA} \bullet \vec{EC} = (-7\mathbf{i} - 2\mathbf{j} - 8\mathbf{k}) \bullet (\mathbf{i} - 10\mathbf{j} - 4\mathbf{k})$$

$$= (-7) \times 1 + (-2) \times (-10) + (-8) \times (-4)$$

$$= -7 + 20 + 32 = 45$$

Magnitude of $\overrightarrow{EA} = \sqrt{(-7)^2 + (-2)^2 + (-8)^2} = 3\sqrt{13}$.

Magnitude of $\overrightarrow{EC} = \sqrt{1^2 + (-10)^2 + (-4)^2} = 3\sqrt{13}$.

$\cos AEC = \dfrac{\overrightarrow{EA} \bullet \overrightarrow{EC}}{|\overrightarrow{EA}||\overrightarrow{EC}|} = \dfrac{45}{3\sqrt{13} \times 3\sqrt{13}} = \dfrac{5}{13}$

$\angle AEC = \cos^{-1}\left(\dfrac{5}{13}\right) = 67.4°$

b M is the mid-point of A and C.

$\overrightarrow{OM} = \dfrac{1}{2}((\mathbf{i} + \mathbf{j} - \mathbf{k}) + (9\mathbf{i} - 7\mathbf{j} + 3\mathbf{k}))$

$\qquad = \dfrac{1}{2}(10\mathbf{i} - 6\mathbf{j} + 2\mathbf{k}) = 5\mathbf{i} - 3\mathbf{j} + \mathbf{k}$

$\overrightarrow{MA} = (\mathbf{i} + \mathbf{j} - \mathbf{k}) - (5\mathbf{i} - 3\mathbf{j} + \mathbf{k}) = -4\mathbf{i} + 4\mathbf{j} - 2\mathbf{k}$

$\overrightarrow{ME} = (8\mathbf{i} + 3\mathbf{j} + 7\mathbf{k}) - (5\mathbf{i} - 3\mathbf{j} + \mathbf{k}) = 3\mathbf{i} + 6\mathbf{j} + 6\mathbf{k}$

$\overrightarrow{MA} \bullet \overrightarrow{ME} = (-4\mathbf{i} + 4\mathbf{j} - 2\mathbf{k}) \bullet (3\mathbf{i} + 6\mathbf{j} + 6\mathbf{k})$

$\qquad = (-4) \times 3 + 4 \times 6 + (-2) \times 6$

$\qquad = -12 + 24 - 12 = 0$

Since $\overrightarrow{MA} \bullet \overrightarrow{ME} = 0$, $\angle AME = 90°$.

> Note that there is no need to find the magnitude of $\overrightarrow{MA}$ or $\overrightarrow{ME}$ in order to demonstrate that the vectors are perpendicular.

c Since ME is perpendicular to the base, the length of ME is the height of the pyramid.

Magnitude of $\overrightarrow{ME} = \sqrt{3^2 + 6^2 + 6^2} = 9$.

The base of the pyramid is a square. Each side is the same length as AB.

$\overrightarrow{AB} = (9\mathbf{i} - \mathbf{j} - 3\mathbf{k}) - (\mathbf{i} + \mathbf{j} - \mathbf{k}) = 8\mathbf{i} - 2\mathbf{j} - 2\mathbf{k}$

Magnitude of $\overrightarrow{AB} = \sqrt{8^2 + (-2)^2 + (-2)^2} = 6\sqrt{2}$.

Area of $ABCD = (6\sqrt{2})^2 = 72$.

Volume of pyramid $= \dfrac{1}{3} \times$ area of base $\times$ height

$\qquad\qquad = \dfrac{1}{3} \times 72 \times 9$

$\qquad\qquad = 216\ \text{unit}^3$

Exercise 7.7B

1 A quadrilateral has vertices $T(2, 5)$, $U(1, 1)$, $V(-3, 2)$ and $W(-2, 6)$.

Show that $TUVW$ is a square.

2 Find the value of a that makes l_1 and l_2 perpendicular.

l_1 $\mathbf{r} = 3\mathbf{i} + 2\mathbf{j} + \mathbf{k} + t(4\mathbf{i} + a\mathbf{j} - 12\mathbf{k})$

l_2 $\mathbf{r} = -2\mathbf{i} + 4\mathbf{j} + 3\mathbf{k} + u(3\mathbf{i} - 2\mathbf{j} - 9\mathbf{k})$

(PS) 3 Which two of the following lines are parallel to each other, which two are perpendicular and which is neither?

l_1 $\mathbf{r} = 5\mathbf{i} + 3\mathbf{j} - 2\mathbf{k} + t(6\mathbf{i} - 3\mathbf{j} + 15\mathbf{k})$

l_2 $\mathbf{r} = -4\mathbf{i} + 2\mathbf{j} - 7\mathbf{k} + t(5\mathbf{i} + 2\mathbf{j} - \mathbf{k})$

l_3 $\mathbf{r} = 9\mathbf{i} + \mathbf{j} - 4\mathbf{k} + t(3\mathbf{i} - \mathbf{j} + 2\mathbf{k})$

l_4 $\mathbf{r} = 6\mathbf{i} - \mathbf{k} + t(4\mathbf{i} - 2\mathbf{j} + 10\mathbf{k})$

l_5 $\mathbf{r} = -2\mathbf{i} + 3\mathbf{j} + 5\mathbf{k} + t(-4\mathbf{i} - 2\mathbf{j} + 5\mathbf{k})$

(C) 4 Given the points $A(-1, -4)$, $B(8, 4)$, $C(3, 6)$, $D(-6, -2)$, $E(-5, -14)$ and $F(4, -6)$, use vectors to show that:

  a $ABCD$ is a parallelogram

  b $CDEF$ is a rhombus

  c $DFBC$ is a trapezium.

(MM) 5 An exhibition is set up so that four of the exhibits are located at the vertices of a square. On a scale model, with the centre of the exhibition at the origin, three of the exhibits are located at $(9, -7, 18)$, $(-6, -1, 28)$ and $(-12, 9, 13)$. Find the coordinates of the fourth exhibit.

(C) 6 A quadrilateral has vertices $K(17, 50, -28)$, $L(32, 20, -18)$, $M(56, 28, -30)$ and $N(41, 58, -40)$.

Show that $KLMN$ is a rectangle.

(C) 7 A quadrilateral has vertices with position vectors $\mathbf{a} = (6\mathbf{i} - 11\mathbf{j} + 4\mathbf{k})$, $\mathbf{b} = (3\mathbf{i} - \mathbf{j} - \mathbf{k})$, $\mathbf{c} = (10\mathbf{i} + \mathbf{j} + 8\mathbf{k})$ and $\mathbf{d} = (23\mathbf{i} - 17\mathbf{j} + 27\mathbf{k})$.

  a Show that $ABCD$ is a kite.

  b Hence find the position vector of the point where the diagonals intersect.

(PS) 8 Points P, Q and R have position vectors $\mathbf{p} = 12\mathbf{i} - \mathbf{j}$, $\mathbf{q} = 6\mathbf{i} + 7\mathbf{j}$ and $\mathbf{r} = -2\mathbf{i} + \mathbf{j}$.

  a Show that QP is perpendicular to QR.

  b Hence find the position vector of the point S such that $PQRS$ is a square.

  c Find the position vector of the centre of the square.

  d Find the area of the square.

9 a Verify that the lines l_1 $\mathbf{r} = \begin{pmatrix} 8 \\ 2 \\ 0 \end{pmatrix} + t\begin{pmatrix} 2 \\ -1 \\ -2 \end{pmatrix}$ and l_2 $\mathbf{r} = \begin{pmatrix} -9 \\ 21 \\ -4 \end{pmatrix} + u\begin{pmatrix} 1 \\ -2 \\ 2 \end{pmatrix}$ are perpendicular and intersect at $(-2, 7, 10)$.

b $ABCD$ is a square with all four vertices on either l_1 or l_2.

Given that A is at $(-5, 13, 4)$, find the coordinates of the other points.

10 A cuboid $ABCDEFGH$ is modelled using a computer such that $\overrightarrow{AB} = (2\mathbf{i} + 4\mathbf{j} + 4\mathbf{k})$, $\overrightarrow{AD} = (-10\mathbf{i} + 10\mathbf{j} - 5\mathbf{k})$ and $\overrightarrow{AE} = (-6\mathbf{i} - 3\mathbf{j} + 6\mathbf{k})$.

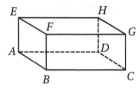

a Given that B has coordinates $(8, 3, -2)$, find the coordinates of H.

b Find $\angle GAC$, in degrees, correct to 1 decimal place.

c Find the angle between the planes $EFGH$ and $EBCH$, correct to the nearest degree.

X is the mid-point of EF.

d Find the cosine of $\angle DXC$.

11 Tetrahedron $OABC$ has vertices $O(0, 0, 0)$, $A(4, 0, 0)$, $B(0, 5, 0)$ and $C(0, 0, 6)$.

a Find the volume of $OABC$.

Tetrahedron $PQRS$ has vertices $P(-1, 4, 2)$, $Q(1, 10, 5)$, $R(11, -2, 6)$ and $S(8, 10, -16)$.

b i Show that PR is perpendicular to PQ.

ii Show that PR is perpendicular to PS.

iii Find the volume of $PQRS$.

12 A cube has vertices $A(5, 3, 1)$, $B(-6, 25, 23)$, $C(16, 47, 12)$, $D(27, p, -10)$, $E(-17, 14, -21)$, $F(-28, 36, 1)$, $G(-6, 58, -10)$ and $H(5, 36, q)$.

a Find the values of p and q.

b Find the position vector of the centre of the cube.

c Find the volume of the cube.

You can use position vectors, the distances between them and the scalar product to identify quadrilaterals. In order to do this, you need to know the properties of different quadrilaterals.

A square	Has four equal sides and four right angles.
A rectangle	Has two pairs of equal sides and four right angles.
A rhombus	Has four equal sides and two pairs of parallel sides.
A parallelogram	Has two pairs of parallel equal sides.
A trapezium	Has one pair of parallel sides.
A kite	Has two pairs of equal adjacent sides.

Example 16

A quadrilateral $RSTU$ has vertices $R(5, 1)$, $S(-1, 8)$, $T(-3, -1)$ and $U(3, -8)$.

Use vectors to show that $RSTU$ is a rhombus.

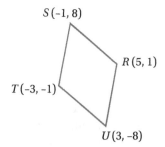

Solution

For $RSTU$ to be a rhombus, you need to prove that:

> • opposite sides RS and UT are the same (parallel and equal magnitude)

> • opposite sides ST and RU are the same (parallel and equal magnitude)

> • adjacent sides such as RS and RU have the same magnitude.

Begin by writing each point as a position vector.

$\mathbf{r} = 5\mathbf{i} + \mathbf{j}$, $\mathbf{s} = -\mathbf{i} + 8\mathbf{j}$, $\mathbf{t} = -3\mathbf{i} - \mathbf{j}$ and $\mathbf{u} = 3\mathbf{i} - 8\mathbf{j}$.

Find the vector $\overrightarrow{RS}$ by subtracting $\mathbf{r}$ from $\mathbf{s}$.

$\overrightarrow{RS} = \mathbf{s} - \mathbf{r} = (-\mathbf{i} + 8\mathbf{j}) - (5\mathbf{i} + \mathbf{j}) = -6\mathbf{i} + 7\mathbf{j}$

Find the vectors of the other sides in the same way.

$\overrightarrow{ST} = \mathbf{t} - \mathbf{s} = (-3\mathbf{i} - \mathbf{j}) - (-\mathbf{i} + 8\mathbf{j}) = -2\mathbf{i} - 9\mathbf{j}$

$\overrightarrow{UT} = \mathbf{t} - \mathbf{u} = (-3\mathbf{i} - \mathbf{j}) - (3\mathbf{i} - 8\mathbf{j}) = -6\mathbf{i} + 7\mathbf{j}$

$\overrightarrow{RU} = \mathbf{u} - \mathbf{r} = (3\mathbf{i} - 8\mathbf{j}) - (5\mathbf{i} + \mathbf{j}) = -2\mathbf{i} - 9\mathbf{j}$

So the vectors $\overrightarrow{RS}$ and $\overrightarrow{UT}$ are the same as are $\overrightarrow{ST}$ and $\overrightarrow{RU}$. Therefore both pairs of opposite sides are equal; they are not only the same length but they are parallel, as required.

All that is left is to ensure that adjacent sides are also the same length.

The formula for the distance between two points is

$d^2 = (x_1 - x_2)^2 + (y_1 - y_2)^2$

For RS:

$d^2 = (5 - (-1))^2 + (1 - 8)^2$

$d^2 = (6)^2 + (-7)^2$

$\quad = 36 + 49$

$\quad = 85$

For *RU*:

$d^2 = (5-3)^2 + (1-(-8))^2$

$\quad = (2)^2 + (9)^2$

$\quad = 4 + 81$

$\quad = 85$

Since d^2 is the same for *RS* and *RU*, the sides are the same length.

All the conditions have been satisfied, so *RSTU* is a rhombus.

> The same result could be obtained by finding vector magnitudes $|\overrightarrow{RS}|$ and $|\overrightarrow{RU}|$.

Stop and think   What other pairs of adjacent sides could be used?

Example 17

Rashmi has drawn a quadrilateral with vertices at $A(100, 165, -234)$, $B(340, 261, -74)$, $C(244, 101, 166)$ and $D(4, 5, 6)$. Show that Rashmi's quadrilateral is a square.

Solution

For *ABCD* to be a square, you need to prove that:

> opposite sides *AB* and *DC* are the same (parallel and equal magnitude)

> opposite sides AD and *BC* are the same (parallel and equal magnitude)

> adjacent sides such as *AB* and *AD* have the same magnitude

> an angle such as *BAD* is a right angle (because then the rest will be too).

You can prove the first three properties in the same way as for **Example 16**. This is left as an exercise for you.

To prove that $\angle BAD$ is a right angle, you can use the scalar product.

Write **a**, **b** and **d** as position vectors. $\mathbf{a} = 100\mathbf{i} + 165\mathbf{j} - 234\mathbf{k}$,
$\mathbf{b} = 340\mathbf{i} + 261\mathbf{j} - 74\mathbf{k}$ and $\mathbf{d} = 4\mathbf{i} + 5\mathbf{j} + 6\mathbf{k}$.

$$\overrightarrow{AB} = \mathbf{b} - \mathbf{a} = (340\mathbf{i} + 261\mathbf{j} - 74\mathbf{k}) - (100\mathbf{i} + 165\mathbf{j} - 234\mathbf{k})$$
$$= 240\mathbf{i} + 96\mathbf{j} + 160\mathbf{k}$$

$$\overrightarrow{AD} = \mathbf{d} - \mathbf{a} = (4\mathbf{i} + 5\mathbf{j} + 6\mathbf{k}) - (100\mathbf{i} + 165\mathbf{j} - 234\mathbf{k})$$
$$= -96\mathbf{i} - 160\mathbf{j} + 240\mathbf{k}$$

$$\overrightarrow{AB} \bullet \overrightarrow{AD} = (240\mathbf{i} + 96\mathbf{j} + 160\mathbf{k}) \bullet (-96\mathbf{i} - 160\mathbf{j} + 240\mathbf{k})$$

$$= 240 \times -96 + 96 \times -160 + 160 \times 240$$

$$= -23\,040 - 15\,360 + 38\,400 = 0$$

Since $\overrightarrow{AB} \bullet \overrightarrow{AD} = 0$, $\angle BAD$ is a right angle.

Mathematics in life and work: Group discussion

A motorway passes a gas station at $(2\mathbf{i} + 9\mathbf{j})$ km heading in the direction of $(-7\mathbf{i} + 2\mathbf{j})$ km.

1 Show that if the motorway continues in the same direction it will pass through the point $(-12\mathbf{i} + 13\mathbf{j})$ km.

2 There is an exit ramp at the point $(-8.5\mathbf{i} + 12\mathbf{j})$ km. The exit ramp is in the direction $(-8\mathbf{i} + \mathbf{j})$ km. Find the angle between the exit ramp and the motorway.

3 What problems would there be if the exit ramp was in the direction $(4\mathbf{i} - 14\mathbf{j})$ km instead?

4 What limitations are there in using the information above for modelling the motorway and exit ramp?

The foot of the perpendicular

Consider a straight line l and a point P not on the line.

The shortest distance between the point and the line can be found by drawing a line through the point that is perpendicular to the line.

The point where the two lines intersect is called the **foot** of the perpendicular from the point to the line, and is the point F.

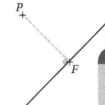

KEY INFORMATION

The shortest distance between a point and a straight line is the length of the perpendicular from the point to the line.

PURE MATHEMATICS 3

Example 18

Line l has equation $\mathbf{r} = 28\mathbf{i} - 10\mathbf{j} - 4\mathbf{k} + t(8\mathbf{i} + 3\mathbf{j} - 6\mathbf{k})$.

The point P lies on the line l such that OP is perpendicular to l.

a Write the position vector of P (in terms of t) in the form $a\mathbf{i} + b\mathbf{j} + c\mathbf{k}$.

b Find the position vector of P.

Solution

a The position vector of P, $\overrightarrow{OP}$, is given by
$(28 + 8t)\mathbf{i} + (-10 + 3t)\mathbf{j} + (-4 - 6t)\mathbf{k}$.

Since you do not know the actual position of P, only that it lies on l, the position vector must be written in terms of t.

b Since OP and l are perpendicular, $\overrightarrow{OP} \bullet (8\mathbf{i} + 3\mathbf{j} - 6\mathbf{k}) = 0$

The vector $\overrightarrow{OP}$ is perpendicular to the direction vector of l.

$$[(28 + 8t)\mathbf{i} + (-10 + 3t)\mathbf{j} + (-4 - 6t)\mathbf{k}] \bullet (8\mathbf{i} + 3\mathbf{j} - 6\mathbf{k}) = 0$$

$$8(28 + 8t) + 3(-10 + 3t) - 6(-4 - 6t) = 0$$

$$224 + 64t - 30 + 9t + 24 + 36t = 0$$

$$109t + 218 = 0$$

$$t = -2$$

Substitute $t = -2$ into $\overrightarrow{OP}$.

Hence $\overrightarrow{OP} = (28 + 8(-2))\mathbf{i} + (-10 + 3(-2))\mathbf{j} + (-4 - 6(-2))\mathbf{k}$.

$\overrightarrow{OP} = 12\mathbf{i} - 16\mathbf{j} + 8\mathbf{k}$

Exercise 7.7C

1 Line l_1 has equation $\mathbf{r} = 16\mathbf{i} + 11\mathbf{j} - 3\mathbf{k} + t(5\mathbf{i} + 7\mathbf{j} - 3\mathbf{k})$.

The point P lies on the line l_1 such that OP is perpendicular to l_1.

Find the position vector of P.

2 Line l has equation $\mathbf{r} = \begin{pmatrix} 3 \\ -25 \\ 13 \end{pmatrix} + t\begin{pmatrix} 4 \\ 5 \\ -1 \end{pmatrix}$.

The point V lies on l such that OV is perpendicular to l.

Find the position vector of V.

3 The line l passes through the points $D(20, 19, 40)$ and $E(-13, 8 -15)$.

a Find the vector equation of l.

Point G has position vector $(-3\mathbf{i} + 8\mathbf{j} + 7\mathbf{k})$.

Point F lies on l such that FG is perpendicular to l.

b Find the position vector of F.

4 The line l_3 has the equation $\mathbf{r} = 4\mathbf{i} + 26\mathbf{j} + 21\mathbf{k} + t(6\mathbf{i} + 5\mathbf{j} + 4\mathbf{k})$.

Point T has position vector $\mathbf{i} + \mathbf{j} - \mathbf{k}$.

The point S lies on the line l_3 such that ST is perpendicular to l_3.

Find the position vector of S.

5 Points $R(27, -17, -1)$ and $S(11, -9, 11)$ lie on the line l_2.

   a Find the equation of the line l_2.

The point Q lies on the line l_2 such that OQ is perpendicular to l_2.

   b Find the position vector of Q.

6 The line l has equation $\mathbf{r} = \begin{pmatrix} -4 \\ 8 \\ -18 \end{pmatrix} + t \begin{pmatrix} 5 \\ 1 \\ 5 \end{pmatrix}$.

   a Show that the point $C(26, 14, 12)$ lies on l.

The point F lies on l such that OF is perpendicular to l.

   b Find the magnitude of $\overrightarrow{OF}$.

   c Find the area of the triangle OFC.

7 Line l passes through the points A and B with position vectors $(17\mathbf{i} - 19\mathbf{j} - 7\mathbf{k})$ and $(-3\mathbf{i} + 16\mathbf{j} + 8\mathbf{k})$.

   a Find the equation of the line l.

Point P lies on the line such that OP is perpendicular to l.

   b Find the position vector of P.

   c Find the ratio $AP : PB$.

   d Find the area of the triangle OPB.

8 The straight line l passes through the points $X(9, 5, -12)$ and $Y(3, 2, -6)$.

Find the shortest distance between the line l and the point $Z(-2, 7, 14)$.

9 Line l_1 has the equation $\mathbf{r_1} = \begin{pmatrix} -5 \\ 7 \\ 1 \end{pmatrix} + t \begin{pmatrix} 3 \\ 1 \\ 4 \end{pmatrix}$.

Line l_2 has the equation $\mathbf{r_2} = \begin{pmatrix} 2 \\ 8 \\ -1 \end{pmatrix} + u \begin{pmatrix} 2 \\ 0 \\ -3 \end{pmatrix}$.

Line l_3 has the equation $\mathbf{r_3} = \begin{pmatrix} 3 \\ 19 \\ 10 \end{pmatrix} + v \begin{pmatrix} -1 \\ 3 \\ 1 \end{pmatrix}$.

l_1 and l_2 intersect at the point T.

The point F lies on l_3 such that TF is perpendicular to l_3.

   a Find the coordinates of the point F.

   b Find the shortest distance between T and l_3.

10 Points $A(-19, -36, -14)$ and $B(21, 24, 10)$ lie on the line l_1.

   a Find the shortest distance between the line l_1 and the point $C(23, -11, 34)$.

   b Hence show that the area of the triangle ABC is 1444.

PURE MATHEMATICS 3

SUMMARY OF KEY POINTS

❭ Vectors can be written in **i j k** notation or as column vectors.

❭ Use Pythagoras' theorem to calculate the magnitude of a vector. In three dimensions you can use the formula $d^2 = (x_1 - x_2)^2 + (y_1 - y_2)^2 + (z_1 - z_2)^2$.

❭ A unit vector is a vector with a magnitude of one.

❭ A vector connecting A and B can be written as $\vec{AB}$ (with a magnitude of $|\vec{AB}|$).

❭ A vector can also be written using lower-case letters such as **a** or $\underset{\sim}{a}$ (with a magnitude of a).

❭ Vectors can be added using the triangle or parallelogram law.

❭ Three points are collinear if they all lie on the same straight line.

❭ The position vector of a point is its position relative to the origin.

❭ To find the displacement vector of B from A, $\vec{AB}$, subtract the position vector of B from the position vector A.

❭ The mid-point of two position vectors is the mean of each of the **i**, **j** and **k** coefficients.

❭ A triangle is:

  ❭ equilateral if all its sides are the same length

  ❭ isosceles if two of its sides are the same length and the third one is a different length

  ❭ scalene if all of its sides are different lengths

  ❭ right-angled if its sides satisfy Pythagoras' theorem.

❭ The vector equation of a line is given by $\mathbf{r} = \mathbf{a} + t\mathbf{b}$, where **a** is the position vector of a point on the line and **b** is the direction of the line.

❭ In two dimensions, lines are either parallel or they intersect.

❭ In three dimensions, lines can be parallel, intersect or be skew to each other.

❭ The angle between two vectors can be found using the scalar product:
$\mathbf{a} \cdot \mathbf{b} = |\mathbf{a}||\mathbf{b}| \cos \theta$.

❭ If $\cos \theta = 0$, then the vectors are perpendicular.

EXAM-STYLE QUESTIONS

1 The points P, Q and R are $(59, -19, 13)$, $(43, 36, -26)$ and $(19, -4, -11)$, respectively.

  a Show that PQR is an isosceles triangle.

  b Show that PQR is a right-angled triangle.

2 The straight line l_1 has equation $\begin{pmatrix} 1 \\ 4 \\ -5 \end{pmatrix} + \lambda \begin{pmatrix} -6 \\ q+5 \\ 3 \end{pmatrix}$. The straight line l_2 has equation $\begin{pmatrix} 1 \\ 4 \\ -5 \end{pmatrix} + \mu \begin{pmatrix} 5 \\ q-6 \\ -4 \end{pmatrix}$.

a Write down the coordinates of the point of intersection of l_1 and l_2.

b Given that l_1 and l_2 are perpendicular, find the possible values of q.

3 The points A and B have position vectors, relative to the origin O. given by $\overrightarrow{OA} = (p+3)\mathbf{i} - 4\mathbf{j}$ and $\overrightarrow{OB} = 8\mathbf{i} + (p-5)\mathbf{j}$

where p is a constant.

a Given that the vectors have the same magnitude, find the value of p.

b Find the acute angle between the vectors.

4 In quadrilateral $OABC$, $\overrightarrow{OA} = 8\mathbf{a}$, $\overrightarrow{OC} = 7\mathbf{c}$ and $\overrightarrow{CB} = 12\mathbf{a}$.

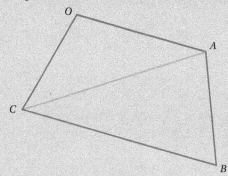

P divides CA in the ratio $3:2$.

a Find the vector $\overrightarrow{OP}$.

b Prove that O, P and B are collinear.

c State the ratio $OP:PB$.

5 The vector $\mathbf{a}$ is given by $\begin{pmatrix} 5 \\ -6 \end{pmatrix}$. The vector $\mathbf{b}$ is given by $\begin{pmatrix} k \\ 1 \end{pmatrix}$.

The vector $\mathbf{c}$ is given by $\mathbf{c} = \mathbf{a} - 2\mathbf{b}$.

a Find in terms of k:

i the vector $\mathbf{c}$

ii the magnitude of the vector $\mathbf{c}$.

b Given that the magnitude of $\mathbf{c}$ is greater than $2\sqrt{17}$, find the set of possible values for k.

6 Points A, B and C are given by $A(5, -2, 4)$, $B(8, 4, 10)$ and $C(14, 7, 4)$.

  a Show that $\overrightarrow{BA}$ is perpendicular to $\overrightarrow{BC}$.

  b Given that $ABCD$ is a square:

    i find the position vector of D

    ii find the position vector of the centre of the square

    iii find the exact area of $ABCD$.

7 The line l_1 passes through $A(2, -5, 1)$ and $B(-2, 3, 2)$. The line l_2 has equation $\mathbf{r} = 19\mathbf{i} - 15\mathbf{j} + 2\mathbf{k} + t(3\mathbf{i} + 2\mathbf{j} + \mathbf{k})$. The line l_3 has equation $\mathbf{r} = -4\mathbf{i} + 10\mathbf{j} - 5\mathbf{k} + u(8\mathbf{i} + a\mathbf{j} - 2\mathbf{k})$. The line l_4 has equation $\mathbf{r} = 7\mathbf{i} - 3\mathbf{j} - 3\mathbf{k} + v(5\mathbf{i} + 3\mathbf{j} - 6\mathbf{k})$.

  a Find the distance between A and B.

  b Find the coordinates of the point at which l_1 and l_2 intersect.

  c Given that l_1 is parallel to l_3, find the value of the constant a.

  d Determine whether l_1 is perpendicular to l_4.

PS 8 The points A, B and C have position vectors, relative to the origin O, given by

$$\overrightarrow{OA} = -8\mathbf{i} - 4\mathbf{j} \qquad \overrightarrow{OB} = -\mathbf{i} + 10\mathbf{j} \qquad \overrightarrow{OC} = 3\mathbf{i} + 8\mathbf{j}$$

Given that $ABCD$ is a rectangle

  a Find the area of $ABCD$.

  b Find the position vector $\overrightarrow{OD}$.

PS 9 In the parallelogram $ABCD$, $\overrightarrow{AB} = 10\mathbf{b}$ and $\overrightarrow{AD} = 10\mathbf{d}$. The point X lies on BC such that $BX:XC = 2:3$ and the point Y lies on DC such that $DY:YC = 3:2$. AD and XY are extended, meeting at the point P.

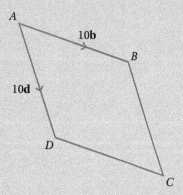

Find the vector $\overrightarrow{BP}$ in terms of $\mathbf{b}$ and $\mathbf{d}$.

10 The straight line l_1 has equation $\begin{pmatrix} -2 \\ -5 \\ 9 \end{pmatrix} + \lambda \begin{pmatrix} -5 \\ 0 \\ 7 \end{pmatrix}$. The straight line l_2 has equation

$\begin{pmatrix} -3 \\ -17 \\ 5 \end{pmatrix} + \mu \begin{pmatrix} 2 \\ 4 \\ -1 \end{pmatrix}$. The lines intersect at the point X.

a Find the coordinates of X.

b Find the acute angle between l_1 and l_2.

11 The points A, B and C are given by $A(1, 4, -5)$, $B(3, 0, 2)$ and $C(-4, 1, 3)$.

a Write down vector equations for the straight line passing through AB and for the straight line passing through AC.

b Given that angle $BAC = \theta$, show that $\cos \theta = \dfrac{58}{7\sqrt{138}}$.

c Find the exact area of triangle ABC.

12 The position vectors of the vertices in parallelogram $CDEF$ are $\mathbf{c} = -4\mathbf{i} + \mathbf{j}$, $\mathbf{d} = -5\mathbf{i} - 6\mathbf{j}$, $\mathbf{e} = 4\mathbf{i} - 3\mathbf{j}$ and $\mathbf{f} = 5\mathbf{i} + 4\mathbf{j}$.

a Find the length of CF, giving the answer in the form $a\sqrt{10}$.

b Show that $\overrightarrow{CD}$ is equal to $\overrightarrow{FE}$.

X is a point on DE such that $DX : XE = 1 : 2$.

c Find the position vector of X.

d Show that XC is perpendicular to XD.

e Find the area of $CDEF$.

f Given that G has position vector $\mathbf{g} = 2\mathbf{i} + 3\mathbf{j}$, show that $DXGC$ is a trapezium.

PS **13** The points A, B, U and V are given by $A(4, 5, -1)$, $B(3, 7, 1)$, $U(13, -1, 5)$ and $V(-3, 7, -11)$.

A, B and W lie on the line l_1. U and V lie on the line l_2.

UVW is an isosceles triangle with a right angle at W.

a Show that l_1 and l_2 are perpendicular.

b Find both possible coordinates of the point W.

14 Points $D(18, 1, -9)$ and $E(10, 3, -3)$ lie on the line l_1.

Points $F(-6, 20, -4)$, $G(8, -29, 31)$ and J lie on the line l_2.

The centre of circle C lies on l_1 such that FJ is a chord of C.

a Show that l_1 and l_2 are perpendicular.

b Hence find the coordinates of J.

PS **15** Find the area of the triangle enclosed by the following lines.

$\mathbf{r} = 3\mathbf{i} + 21\mathbf{j} - 7\mathbf{k} + t(2\mathbf{i} - 2\mathbf{j} + 3\mathbf{k})$

$\mathbf{r} = -\mathbf{i} - 7\mathbf{j} + 3\mathbf{k} + u(2\mathbf{i} + 6\mathbf{j} - \mathbf{k})$

$\mathbf{r} = 17\mathbf{i} + 19\mathbf{j} + 8\mathbf{k} + v(2\mathbf{i} + 2\mathbf{j} + \mathbf{k})$

16 Line l has equation $\mathbf{r} = 28\mathbf{i} + 20\mathbf{j} + 4\mathbf{k} + t(3\mathbf{i} - \mathbf{j} + 3\mathbf{k})$.

The point R lies on the line l such that OR is perpendicular to l.

a Find the position vector of R.

b Hence find the shortest distance between O and l.

17 Line l_1 passes through the points $Q(-2, 2, 0)$ and $R(0, -2, -6)$.

Line l_2 has the vector equation $\mathbf{r} = -40\mathbf{i} - 2\mathbf{j} - \mathbf{k} + t(7\mathbf{i} + 2\mathbf{j} + 2\mathbf{k})$.

a Find a vector equation for the line l_1.

b Show that $S(23, 16, 17)$ lies on the line l_2.

c Find the position vector of the point of intersection, P, of lines l_1 and l_2.

d Find the angle RPS, correct to 2 decimal places.

e Find the area of the triangle PRS correct to 3 significant figures.

18 Tetrahedron $ABCD$ has vertices $A(4, 3, -1)$, $B(-4, 5, 2)$, $C(6, -1, 0)$ and $D(10, 11, 19)$.

a Find the area of triangle ABC in the form $a\sqrt{6}$.

The point $E(1, 2, 1)$ lies in the same plane as the triangle ABC.

b Show that $\angle AED = 90°$.

c Find the volume of $ABCD$.

19 $G(7, -5, 6)$ and $H(15, 11, 4)$ lie on the line l_1.

a Show that GH is 18 units in length.

Line l_2 is parallel to line l_1 and passes through the point $E(7, -23, 24)$.

b Show that the point $F(23, 9, 20)$ lies on the line l_2.

c Show that angle FEG is 45°.

d Show that l_1 is perpendicular to FH.

e Show that the trapezium $EFHG$ has an area of 486 units^2.

(PS) 20 Points $A(-7, -4, 9)$ and $B(8, 5, 3)$ lie on the line l_1.

a Find a vector equation for l_1.

The line l_2 has the vector equation $\mathbf{r} = \begin{pmatrix} 6 \\ 11 \\ 7 \end{pmatrix} + t\begin{pmatrix} -1 \\ 3 \\ 2 \end{pmatrix}$ where C has coordinates $(6, 11, 7)$.

b Show that the point B lies on the line l_2.

c Show that l_1 and l_2 are perpendicular.

d Find the angle CAB, in degrees, correct to 1 decimal place.

e Find the area of the triangle ABC.

The point D is the reflection of the point C in the line l_1.

f Find the coordinates of D.

C 21 The line l is given by the equation $\mathbf{r} = \begin{pmatrix} -19 \\ 14 \\ -5 \end{pmatrix} + t \begin{pmatrix} 1 \\ -3 \\ a \end{pmatrix}$.

Point T has coordinates $(-2, 5, 8)$.

The point F lies on the line l such that $\overrightarrow{TF}$ is perpendicular to l.

a Show that $t = \dfrac{13a + 44}{a^2 + 10}$.

b Given that $t = 5$, find both possible position vectors for F.

Mathematics in life and work

A main road passes by the village of Patrycja. According to a cartographer's model, Patrycja is located at $P(16.4\mathbf{i} + 23.1\mathbf{j} + 0.5\mathbf{k})$ km. The main road leaves a motorway at $M(-2.2\mathbf{i} + 21.5\mathbf{j} + 0.7\mathbf{k})$ km and is straight with a direction of $(13\mathbf{i} - 14\mathbf{j} - \mathbf{k})$ km.

1 Find the vector $\overrightarrow{MP}$.

2 Hence find the angle between MP and the main road, correct to 1 decimal place.

3 Hence find the shortest distance between Patrycja and the main road, correct to the nearest kilometre.

4 Find the point on the main road which is closest to Patrycja.

8 DIFFERENTIAL EQUATIONS

Mathematics in life and work

People have always been faced with the challenge of understanding and modelling an ever-changing world. The mathematics of change is studied using calculus, and specifically differential equations, a technique devised by Newton and Leibnitz in the 17th century. Differential equations are very common in all branches of science and engineering, where they can illustrate the relationship between variables that change over time or space. The solution of a differential equation is a formula that gives one variable in terms of another.

This kind of mathematics is used in a great variety of careers – for example:

> If you were an economist, you would use differential equations to find optimum investment strategies. Economists have to consider a wide variety of changes (such as the current economic system, the international economic climate and world events, among many other variables) in order to try to predict what will happen to the markets.

> If you were a physicist, you might use differential equations to describe the motion of waves, pendulums or chaotic systems.

> If you were an engineer modelling the heat loss of buildings, there is a clear model for the rate of change of temperature that uses differential equations.

> If you were a forensic scientist, you may use the same model by measuring the temperature of a body, and comparing it with the surrounding temperature.

In this chapter you will consider Newton's law of cooling, which states that the rate at which the temperature of a body (for example, a building or a cup of tea) falls is proportional to the difference between the temperature of the object and the surrounding temperature. An important design consideration for new buildings is insulation, which will help the building to stay cool in a warm environment or keep warm in a cool environment.

If a building has been heated up to a satisfactory level then the insulation can be tested by turning off the heat source and noting how the temperature changes. This graph shows what may happen to the temperature over time.

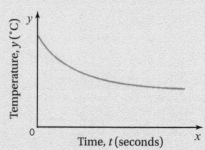

However, the graph will change in different circumstances. For example, if the building has thin walls or lots of windows, it may cool more quickly than if it has a thick layer of insulation.

LEARNING OBJECTIVES

You will learn how to:

› formulate a simple statement involving a rate of change as a differential equation

› use integration to find a general form of solution to a first-order differential equation

› use an initial condition to find a particular solution

› use the separation of variables method to find a general form of solution for a first order differential equation

› interpret the solution of a differential equation in the context of a problem being modelled by the equation.

LANGUAGE OF MATHEMATICS

Key words and phrases you will meet in this chapter:

› constant of proportionality, differential equation, exponential decay, exponential growth, general solution, particular solution, rate of change, separation of variables

PREREQUISITE KNOWLEDGE

You should already know how to:

› integrate simple exponential and trigonometric expressions (as seen in **Chapter 5 Integration**)

› manipulate trigonometric expressions (see **Chapter 3 Trigonometry**) and other algebraic expressions

› understand the definition and properties of e^x and $\ln x$, including their relationship as inverse functions and their graphs.

You should be able to complete the following questions correctly:

1 Differentiate the following with respect to x.

   a $\sin(x^2 + 1)$ **b** $e^{-x}\cos 2x$ **c** $\ln(3x + 2)$

2 Find these integrals

   a $\int \dfrac{4}{\sqrt{x}}\,dx$ **b** $\int e^{4x}\,dx$ **c** $\int \sec^2 x\,dx$

   d $\int \dfrac{3}{3x + 2}\,dx$ **e** $\int \dfrac{6x^2}{2x^3 + 4}\,dx$ **f** $\int \sin(4x)\,dx$

3 Simplify these expressions.

   a $4\cos x \times \sin x$ **b** $\sin^2 x - \cos^2 x$ **c** $e^x \div e^{-x}$

PURE MATHEMATICS 3

8.1 Constructing differential equations

In order to solve a **differential equation**, it may be necessary first to construct the equation itself. If there is a link between how the variable x is changing as time (t) changes, then this is the **rate of change** and can be written as $\frac{dx}{dt}$.

If the gradient of a line varies according to a rule or law then you will have the differential equation for $\frac{dy}{dx}$.

The solution to the differential equation is such that when it is differentiated and substituted into the original equation it will hold true. The process of differentiating will also remove any constant terms.

A **general solution** is one in which the constants are not known, and they are usually written as '$+ c$'. You may be given extra information to calculate the value of this constant. The solution is then called the **particular solution**.

> **KEY INFORMATION**
>
> The rate of change of x with respect to time (t), $\frac{dx}{dt}$, is sometimes written as $\dot{x}$.
>
> Similarly, $\frac{d^2x}{dt^2} = \ddot{x}$.

If $y = c$, then $\frac{dy}{dx} = 0$.

Example 1

A culture of bacteria is modelled as growing at a rate that is proportional to its size. The population size of the culture at time t hours is B. Initially, B is 200, and after 10 hours it has become 400.

a Use this information to form a differential equation.

b Show that $B = ae^{kt}$ is the general solution, and find the particular solution.

c Describe any limitations of this solution.

This tells you that the population size is doubling every 10 hours.

Solution

a You are told that the rate of change of the size of the culture is proportional to its size.

$$\frac{dB}{dt} \propto B$$

So you can introduce a **constant of proportionality**, k.

$$\frac{dB}{dt} = kB$$

This is the differential equation for the growth of bacteria in the model.

If two variables vary according to a common ratio or are in proportion to each other, then you can use a multiplying factor, called the constant of proportionality.

b You can check that $B = ae^{kt}$ is a solution by differentiation.

$$\frac{dB}{dt} = ake^{kt}$$

$$= k \times ae^{kt} = kB$$

This shows that $B = ae^{kt}$ is a solution.

Now you can substitute $B = 200$ at $t = 0$. •——

$200 = ae^0$ so $a = 200$ •——

You also know that at $t = 10$, $B = 400$.

$400 = 200e^{10k}$

Rearrange.

$e^{10k} = 2$

$10k = \ln 2$

$k = \dfrac{\ln 2}{10} = 0.069$

So the particular solution is $B = 200e^{0.069t}$.

> $B = 200$ is the initial condition.

> At $t = 0$, e^{-kt} becomes e^0 which is equal to 1.

c As t becomes large, the size of the culture will become very large, and the model will no longer be valid.

Stop and think
The growth of bacteria is a concern in, for example, hospitals, restaurants and farms. Ensuring hygiene is crucial. Mathematical models can be improved to allow for many variables. Which variables in each of these locations do you think might affect the growth of bacteria the most? How might this be mathematically described?

The solution to **Example 1 part b** is of the form $x = Ae^{kt}$, $k > 0$. This model is called **exponential growth**.

Consider the situation at the start of this chapter relating to the temperature loss of a building. This is an example of **exponential decay**.

Exercise 8.1A

1 Given that $\dfrac{dy}{dx} = 4x^3 + 4x + d$, show that $y = x^4 + 2x^2 + 2x + 4$ is a particular solution of this differential equation, and state the value of d.

2 Given that $\dfrac{dy}{dx} = k\sqrt{x}$, show that $y = x^{1.5} + c$ is a general solution, and state the value of k.

3 The gradient at any point on the graph of $y = f(x)$ is proportional to the square of the x-coordinate.

a Write down a differential equation to express this relationship.

b Show that $y = x^3 + 4$ is a solution of your differential equation.

Ⓒ Communication   ⓂⓂ Mathematical modelling   ⓅⓈ Problem solving   **227**

4 The gradient at any point on the graph of $y = f(x)$ is inversely proportional to the y-coordinate. State the value of the constant of proportionality in each of **parts b** and **c**.

   a Write down a differential equation to express this relationship.

   b Show that $y = \sqrt{x}$ is a solution of your differential equation.

   c Show that $y = a\sqrt{x}$ is a solution of your differential equation for any value of a.

5 The rate of increase of the population of a country at any time is equal to 0.5% of the population size at that time.

   a Given that y is the population in millions and t is time in years, write down a differential equation to show this fact.

   b Show that a solution of the equation is $y = ae^{0.005t}$, where a is a constant.

   c The population now is 50 million. Estimate the population in 20 years' time.

6 The gradient at any point on the graph of $y = f(x)$ is equal to the product of the two coordinates.

   a Write down a differential equation to express this relationship.

   b Show that $y = e^{\frac{1}{2}x^2}$ is a solution of your differential equation and state the constant of proportionality of the relationship between $f'(x)$ and the product of the two coordinates.

7 A particle is moving away from point A.

The speed of the particle is inversely proportional to the distance, x m, from A.

When the distance is 5 m, the speed is $0.4 \, \text{m s}^{-1}$.

   a Show that $x\dfrac{dx}{dt} = 2$.

   b Find the acceleration of the particle when the distance is 5 m.

   c What is the long-term motion of the particle?

8 A spherical balloon is being inflated by pumping in air at a constant rate.

The rate of increase of the radius of the sphere with time is inversely proportional to the square of the radius.

   a Write down a differential equation to show this. Use r for the radius in cm and t for the time in seconds.

When the radius is 10 cm, the rate of increase of the radius is $0.5 \, \text{cm s}^{-1}$.

   b Show that the differential equation can be written as $r^2\dfrac{dr}{dt} = 50$.

   c Show that $r = a\sqrt[3]{t}$ is a solution to the differential equation where $a = \sqrt[3]{150}$.

9 As a snowball rolls down a slope it speeds up according to how far from the top of the hill it is. Its speed is proportional to the cube root of the distance.

   a At a distance of 8 m the speed is $4 \, \text{ms}^{-1}$. Write down a differential equation linking speed and distance, x m, clearly stating the constant of proportionality.

   b Show that $x = \left(\dfrac{2t}{3} + 4\right)^{\frac{3}{2}}$ is a valid solution. State the initial value of x that this solution implies.

   c Explain why $x = 0$ at $t = 0$ is not a valid initial situation.

10 The rate of growth of population of a bacteria, $\frac{dP}{dt}$, in a laboratory is inversely proportional to the size of the population, P thousand after t minutes.

   a Form a general differential equation and show that $P = \sqrt{2kt + c}$ is a solution.

   b If at $t = 0$, $P = 2$ and $\frac{dP}{dt} = 2$, find the population and rate of change of population after 2 minutes.

   c Sketch the solution to show the long-term pattern of growth. Comment on the shape.

8.2 Solving differential equations by integration

Here are three differential equations:

$$\frac{dy}{dt} = -10e^{-0.2t}$$

$$\frac{dy}{dx} = \frac{2}{x^2 + 1}$$

$$\frac{dx}{dt} = \frac{1}{6x^2}.$$

You can check whether a particular function is a solution by differentiating it, but finding a general solution requires integration.

Look at each of these differential equations in turn.

> **KEY INFORMATION**
>
> Generally, if $\frac{dy}{dx} = f(x)$, then
>
> $y = \int f(x) \, dx$.

Method 1: $\frac{dy}{dt} = -10e^{-0.2t}$

In this case $\frac{dy}{dt}$ is a function of t, so you can simply integrate it:

$$y = \int -10e^{-0.2t} \, dt.$$

You could guess that the answer is a multiple of $e^{-0.2t}$.

$$\frac{d}{dx}\left(e^{-0.2t}\right) = -0.2e^{-0.2t}$$

$$\int -10e^{-0.2t} \, dt = \frac{-10}{-0.2}e^{-0.2t} + c. \bullet\!-\!-\!-$$

This gives a general solution of $y = 50e^{-0.2t} + c. \bullet\!-\!-\!-$

Method 2: $\frac{dy}{dx} = \frac{2}{x^2 + 1}$

In this case, you have a function of x, so can integrate both sides with respect to x to give

$$y = \int \frac{2}{x^2 + 1} \, dx.$$

This is looks like the standard result, $\frac{d}{dx}(\tan^{-1} x) = \frac{1}{x^2 + 1}$.

> Notice that there is an arbitrary constant in the answer. Because solving a differential equation involves integration, there will always be an arbitrary constant in the general solution.

> You can check by differentiation that this is a solution to the differential equation.

> The general solution is one in which the arbitrary constant is not known. More information is needed in order to find the value of this constant, such as the initial conditions.

PURE MATHEMATICS 3

So $\int \frac{2}{x^2+1} \, dx = 2\tan^{-1}x + c.$

Which gives a general solution of $y = 2\tan^{-1}x + c.$

Method 3: $\frac{dx}{dt} = \frac{1}{6x^2}$

This may look similar to the example above, but in this case $\frac{dx}{dt}$ is a function of x and does not involve t.

You know that the reciprocal of $\frac{dx}{dt}$ is $\frac{dt}{dx}$ so you can write

$$\frac{dt}{dx} = 1 \div \left(\frac{1}{6x^2}\right) = 6x^2$$

Now $\frac{dt}{dx}$ is a function of x and not t, so you can integrate.

$$t = \int 6x^2 \, dx = 2x^3 + c$$

$$= 2x^3 + c$$

So $x = \sqrt[3]{\frac{t-c}{2}}.$

A common type of differential equation occurs when the rate of change of a variable is proportional to the value of the variable.

The equation will be of the form $\frac{dx}{dt} = kx$ where k is a constant.

This differential equation could have also been solved using a method called separation of variables, which is covered later in this chapter.

The final solution is given as a function of t.

Mathematics in life and work: Group discussion

You have been asked to find out how effective a building's insulation is in cool environments. The building is designed to be used at 20 °C, maintaining this room temperature for as long as possible. If the room falls below 15 °C, then the room is no longer fit to work in. Experimental data tells you that after 20 minutes, the temperature of the room has fallen to 18 °C.

1   What are the different variables in this question? How might you determine what is an acceptable time for the room to stay at an appropriate temperature?

2   A simple model suggests a linear rate of decline in temperature. We can write this as

$$\frac{dx}{dt} = -k.$$

What do the letters x, t and k represent? Solve this differential equation. Which of the numbers in the initial set-up are used?

3   What does this model suggest about the temperature of the room at different times? How might you adjust the model to account for the building being in a different environment? Discuss the limitations of the model. At what point does it become an unrealistic model? How should the model be changed if the building was to be located in a warm environment? What would the graph look like in this scenario?

Example 2

The number of fish in a lake is increasing annually. The population, x fish, in t years' time, is modelled by the equation $\dfrac{\mathrm{d}x}{\mathrm{d}t} = 0.2x$.

This year there are 300 fish.

a Work out a formula for the number of fish after t years.

b Describe any limitations in this solution.

Solution

a You can write the equation as $\dfrac{\mathrm{d}t}{\mathrm{d}x} = \dfrac{5}{x}$.

Integrate: $t = \displaystyle\int \dfrac{5}{x}\,\mathrm{d}x = 5\ln x + c$, where c is a constant.

$\ln x = \dfrac{t - c}{5}$

$x = \mathrm{e}^{0.2(t - c)}$

> A modulus sign on x is not necessary because it must be positive due to the context in the example.

You could leave the solution as it is, but is more usual to tidy it up, and write it as:

$$x = \mathrm{e}^{0.2t - 0.2c} = \mathrm{e}^{0.2t} \times \mathrm{e}^{-0.2c}.$$

You can write this as $x = k\mathrm{e}^{0.2t}$, where you now have a different constant, k.

When $t = 0$, $x = 300$, so $300 = k \times 1$ and therefore $k = 300$.

The formula is $x = 300\mathrm{e}^{0.2t}$.

b This model implies that the population increases by so there are more and more fish each year. This cannot continue for a long period, because of restrictions on space and food in the lake.

The solution will then no longer be valid.

Example 3

A stone is dropped from rest, from the top of a high building.

The speed $v\,\mathrm{m\,s}^{-1}$ after t seconds is given by the differential equation $\dfrac{\mathrm{d}v}{\mathrm{d}t} = 10 - 0.5v$.

a Find an expression for v in terms of t.

b Show that the speed approaches a limiting value.

c Explain any limitations to this solution.

Solution

a You can write $\dfrac{dt}{dv} = \dfrac{1}{10 - 0.5v}$.

> This step is necessary in order to be able to integrate with respect to v.

Then $t = \int \dfrac{1}{10 - 0.5v}\, dv$.

> Making the substitution $u = 10 - 0.5v$ would be another way to find the solution.

It looks as if the solution is a multiple of $\ln|10 - 0.5v|$.

Use the chain rule.

$$\dfrac{d}{dv}\left(\ln|10 - 0.5v|\right) = \dfrac{1}{10 - 0.5v} \times -0.5$$

So $t = \int \dfrac{1}{10 - 0.5v}\, dv = -2\ln|10 - 0.5v| + c$.

> Recall the result
> $$\int \dfrac{f'(x)}{f(x)}\, dx = \ln|f(x)| + c.$$

Rearrange.

$$-2\ln|10 - 0.5v| = t - c$$

$$\ln|10 - 0.5v| = -\frac{1}{2}(t - c)$$

$$10 - 0.5v = e^{-\frac{1}{2}(t-c)}$$

$$0.5v = 10 - e^{-\frac{1}{2}(t-c)}$$

Multiply by 2.

$$v = 20 - 2e^{-\frac{1}{2}(t-c)}$$

Write $2e^{-\frac{1}{2}(t-c)} = 2e^{-0.5t} \times e^{0.5c} = Ae^{-0.5t}$, where the arbitrary constant is now $A = 2e^{0.5c}$.

> You can put the constant in any convenient form.

So $v = 20 - Ae^{-0.5t}$.

This is the general solution.

For the particular solution in this case, you can use the information hidden in the question to substitute values. The question states that the stone starts from rest, so it has an initial condition of $v = 0$ when $t = 0$. Substitute these into the general solution.

$0 = 20 - A$ so $A = 20$.

Therefore $v = 20 - 20e^{-0.5t}$.

> This is the particular solution required, in this case.

b As t increases, $e^{-0.5t} \to 0$.
Therefore the speed approaches a limiting value of $20 - 0 = 20\,\mathrm{m\,s^{-1}}$.

c The solution is only valid while the stone is in the air. If the building is not very high, then the stone could hit the ground before it gets close to the limiting value of $20\,\mathrm{m\,s^{-1}}$.

> **Stop and think**
>
> This situation is an improvement on the model for freefall, which ignores air resistance, as this model will tend towards a terminal velocity. The model can be adapted to study the motion of cars travelling at high speeds on a motorway. What other situations are similar to the motion of the stone?

Exercise 8.2A

1 Find the particular solution to each of these differential equations.

    a $\dfrac{dy}{dx} = 2 - 0.3x^2$ given that $y = 2$ when $x = 2$.

    b $\dfrac{dy}{dx} = \cos(3x)$ given that $y = 4$ when $x = \dfrac{\pi}{3}$.

    c $\dfrac{dy}{dx} = \dfrac{2 + x^2 - 4x^5}{x}$ given that $y = 2$ when $x = 1$.

2 Solve the differential equation $\dfrac{dy}{dx} = 8ax^3$.

3 Find the general solution of the differential equation $\dfrac{dy}{dx} = \sqrt{y}$.

4 Solve the differential equation $\dfrac{dy}{dx} = 0.1y^2$.

(PS) 5 Find the equation of the curve that passes through the point $(1, 9)$ and has a gradient function of $8x + \dfrac{6}{x^3}$.

(PS) 6 A particle, which initially is at the origin, moves along a straight line such that the rate of change of its position follows the rule $\dot{x} = t^3 - 6t^2 + at$, at time t seconds.

    a Solve the differentiatial equation to find the displacement, x metres, of the particle in terms of a.

    b Find an expression for the acceleration, $\ddot{x}$ of the particle in terms of a.

    c What is the range of values of a for which the particle does not return to the origin?

(C) 7 There are initially 100 000 bacteria in a colony in a laboratory.

    The number of bacteria, y, after t days is modelled by the differential equation $\dfrac{dy}{dt} = \dfrac{1}{2}y^{0.8}$.

    a Find a formula for the number of bacteria after t days.

    b Find the number of bacteria after one week.

    c Comment on any limitations on the validity of the solution.

8 A particle, which is initially at $x = \pi$, moves along a straight line such that $\dot{x} = \cos^2 x$.

    a Solve the differential equation to find the displacement, x metres, of the particle.

    b What is the limiting value of x as t increases?

9 A car accelerates from rest.

The speed v m s^{-1} after t s is given by the differential equation $\dfrac{dv}{dt} = 10 - 0.5v$.

a Find an expression for v in terms of t.

b Show that the speed approaches a limiting value.

c Comment on the limitations of the model.

10 By use of a standard result, find an expression for x in terms of t, given that

$$\frac{dx}{dt} = \frac{\sqrt{1 - x^2}}{-4}.$$

Sketch the solution, given $x = 1$ at $t = 0$.

8.3 Separation of variables

Up until now, you have only dealt with questions of the form $\dfrac{dy}{dx} = f(x)$. But sometimes a situation arises where you have a function involving both x and y, so simply integrating will not work.

The method is to imagine that $\dfrac{dy}{dx}$ is not a single expression, but that dy and dx can be written separately, treating the expression as a fraction. Remember that the expression $\dfrac{dy}{dx}$ is the limit of the fraction $\dfrac{\delta y}{\delta x}$ as $\delta x \to 0$ and $\delta y \to 0$. It is then possible to rearrange the expression so that all the x terms are on one side of the equals sign, and all the y terms are on the other side. This method is called **separation of variables**.

Example 4

Solve the differential equation $\dfrac{dy}{dx} = -xy^2$.

Solution

Multiply both sides by dx and divide both sides by $-y^2$.

$$-\frac{1}{y^2}\,dy = x\,dx$$

For this expression to be mathematically valid, you must now integrate both sides.

$$\int -\frac{1}{y^2}\,dy = \int x\,dx$$

$$\frac{1}{y} = \frac{1}{2}x^2 + c$$

KEY INFORMATION

Generally, if $\dfrac{dy}{dx} = f(x)\,g(y)$, then $\displaystyle\int \frac{1}{g(y)}\,dy = \int f(x)\,dx$.

For this step, it helps to rewrite $-\dfrac{1}{y^2}$ as $-y^{-2}$.

Don't forget the arbitrary constant – but you only need one, not two.

You could leave the answer in this implicit form or you could rearrange it as

$$y = \frac{1}{\frac{1}{2}x^2 + c} \quad \text{or} \quad y = \frac{2}{x^2 + 2c}.$$

You might also replace $2c$ with a different arbitrary constant,

k, say, and write $y = \dfrac{2}{x^2 + k}$.

Example 5

Rose is cleaning her fish tank. To do this, she stirs the water in the tank whilst pumping clean water into it at a rate of 10 litres per minute and draining out the mixture at a rate of 15 litres per minute. The tank holds 200 litres of water, and initially she estimates there to be 1 kg of dirt mixed in with the water in the tank.

a If there is x kg of dirt in the tank at time t minutes, then show that the rate of change of dirt in the tank is given by the differential equation
$$\frac{dx}{dt} = \frac{-3x}{40 - t}.$$

b Solve this equation. How much dirt is in the tank after 10 minutes, and how full will the tank be?

c Describe how effective this method is for cleaning the fish tank. Discuss what changes you might suggest to the technique.

Solution

a First consider that the volume of liquid in the tank will be $V = 200 - 5t$ after t minutes.

> $10t$ is the volume of water pumped into the tank after t minutes. $15t$ is the volume pumped out, so overall change will be $10t - 15t$, which is $-5t$.

The concentration of the dirt after t minutes will therefore be

$$\frac{x}{V} = \frac{x}{200 - 5t}.$$

The rate of change of x is this multiplied by the rate at which the liquid is leaving the tank.

$$\frac{dx}{dt} = -\frac{x}{200 - 5t} \times 15 = \frac{-15x}{200 - 5t}$$

which simplifies to

$$\frac{dx}{dt} = \frac{-3x}{40 - t}.$$

b To solve this you need to separate the variables, which gives

$$\int \frac{1}{x} \, dx = \int \frac{-3}{40 - t} \, dt.$$

This can be solved using integration by substitution, giving

$$\ln (x) = 3 \ln (40 - t) + c$$

$$= \ln (40 - t)^3 + c.$$

$x = A(40 - t)^3$ – this is the general solution.

> The arbitrary constant is now $A = e^c$.

Initially you know that $x = 1 \, \text{kg}$, so

$$1 = A \times (40)^3$$

$$A = \frac{1}{(40)^3} = \frac{1}{64\,000}.$$

which gives the particular solution

$$x = \frac{(40 - t)^3}{64\,000}$$

Using this equation, at $t = 10$,

$$x = \frac{(40 - 10)^3}{64\,000} = 0.422 \, \text{kg} \, (3 \, \text{s.f.}).$$

The amount of liquid is $200 - 5t$.

At $t = 10$ this gives $200 - 5 \times 10 = 150 \, \text{litres}$.

c After 10 minutes, over half of the dirt was removed. However, at $t = 20$ minutes there is 0.125 kg left, so the rate at which the dirt is being removed is decreasing. Rose may want to stop the process after a certain level, for example, being satisfied with the dirt level being halved. Solving $200 - 5t = 0$ gives $t = 40$. This means that after 40 minutes the tank will have no water left in it! Rose could change the method, by aiming to remove the same volume that she is adding, so that the volume in the tank remains constant.

> **Stop and think**
>
> The mathematics of cleaning the fish tank in this manner is similar to the medical procedure dialysis, used to remove waste products from the blood. The process involves removing the patient's blood at a fixed rate, and returning it at the same rate. How is this process the same as the question above? How does the model need to be changed?

Mathematics in life and work: Group discussion

Earlier in the chapter you considered the use of insulation in building design. If you were involved as an architect or civil engineer in the design, you would want to be able to model the changes accurately. Initially the room is at $20\,^\circ\text{C}$. The model for the cooling of the building is now refined to take into account Newton's law of cooling, so the improved differential equation is $\frac{dx}{dt} = -k(x - a)$.

1. What do the letters x, t, k and a represent? What factors will change the value of k? If k increases, does this make the building insulation more or less effective?

2. What is the general solution to the differential equation? Write your answer with x as the subject of the formula.

3. In many colder countries there are legal requirements to keep work spaces above $15\,^\circ\text{C}$. You require the room to remain at this temperature for at least 40 minutes after the heating is turned off. Use the conditions that the external temperature is $5\,^\circ\text{C}$ and after 10 minutes the room is at $19\,^\circ\text{C}$. Find the constants in this situation. Does the building keep warm enough for long enough?

4. The building design is now tested in a colder environment, where the external temperature is $-5\,^\circ\text{C}$. After 10 minutes its temperature is $18\,^\circ\text{C}$. Update the variables in the solution. Does the room manage to keep above the legal temperature for at least 40 minutes?

Exercise 8.3A

1. State whether these differential equations can be solved by using the separation of variables method, and find the general solution of those that can be.

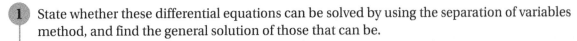

   a $\frac{dy}{dx} = 3x^2y$

   b $\frac{dy}{dx} = 3x\sqrt{y}$

   c $\frac{dy}{dx} = 3x + y$

   d $\frac{dy}{dx} = e^{x+y}$

   e $\frac{dy}{dx} + \frac{3x}{y^2} = 0$

   f $\frac{dy}{dx} = \cos(x) + \sin(y)$

2. Find the particular solution to $\frac{dy}{dx} = (3x^2 + 4)y^2$, given that $y = -0.1$ when $x = 2$.

3. **a** Solve the differential equation $\frac{dy}{dx} + \frac{y}{x} = 0$.

   b A curve passes through the point (6, 4) and $\frac{dy}{dx} + \frac{y}{x} = 0$.
   Find the equation of the curve.

 4 The equation of an ellipse satisfies the differential equation $\frac{dy}{dx} = -\frac{2x}{3y}$.

   a Show that the equation of the curve can be written as $x^2 + ny^2 = a$, where a is an arbitrary constant and n is a number that you must find.

   The point (5, 4) is on the curve.

   b Find the coordinates of the points where the ellipse crosses the x-axis.

5 Find the general solution to the differential equation $\dfrac{dy}{dx} = \dfrac{y-3}{y}$.

6 Find the general solution to the differential equation $\dfrac{dy}{dx} = \dfrac{y^2-4}{x}$.

7 A radioactive substance is decaying.

(PS)

The mass x mg left after t days satisfies the differential equation $\dfrac{dx}{dt} = -\lambda x$, where $\lambda > 0$.

a Show that a general solution can be written in the form $x = ae^{-\lambda t}$ and explain the interpretation of the constant a.

The half-life of a radioactive element is the time for half of the substance to decay.

b Show that the half-life of this element is $\dfrac{\ln 2}{\lambda}$.

8 Kirchhoff's law shows the relationship between current (I, amps), voltage (V, volts), resistance (R, ohms), and an electrical inductance (L, henrys).

$$V = L\dfrac{dI}{dt} + IR$$

Given that $V = 50$ volts, $L = 4$ henrys and $R = 3$ ohms, solve the differential equation to find an expression for I in terms of t.

(PS) **9** The gradient of a curve at any point is $y\cos x$. At $x = 0$, $y = 1$ and $\dfrac{dy}{dx} = 0$.

Find the equation of the curve in the form $y = f(x)$.

Sketch the curve for $-10 < x < 10$.

Prove that the curve does not intersect with the curve $y = \sin x$.

10 a Given that $\dfrac{dy}{dx} = \dfrac{2\sin^2 x}{y}$, find an expression for y in terms of x.

The line passes through $(0, 0)$

b Sketch the particular solution and the curve $y = \sqrt{2x}$ on the same axis.

SUMMARY OF KEY POINTS

> A differential equation is an equation that has differential terms, such as $\frac{dy}{dx}$.

> A general solution involves one or more arbitrary constant.

> A particular solution is a solution to a differential equation in which all the constants are known.

> Separation of variables is a technique that ensures that the variables, usually x and y, are on opposite sides of the equals sign.

> The rate of change of x with respect to time, t, is written as $\frac{dx}{dt}$ or $\dot{x}$.

> If the gradient of a line varies according to a rule or law, then you will have the differential equation for $\frac{dy}{dx}$.

> Generally, if $\frac{dy}{dx} = f(x)$, then $y = \int f(x)\,dx$.

> Generally, if $\frac{dy}{dx} = f(y)$, then this rearranges to $\frac{dx}{dy} = \frac{1}{f(y)} \Rightarrow = \int \frac{1}{f(y)}\,dy$.

> Generally, if $\frac{dy}{dx} = f(x)g(y)$, then $\int \frac{1}{g(y)}\,dy = \int f(x)\,dx$.

EXAM-STYLE QUESTIONS

1   **a**   Given that $x = \frac{4}{t}$, show that $\frac{dx}{dt} = \frac{-x^2}{4}$.

     b   Find the value of $\frac{dx}{dt}$ when t = 3.

2   **a**   Given that $s = 15\cos^2(4t)$, find an expression for $\frac{ds}{dt}$.

     The variables x and y are related by the differential equation $\frac{dy}{dx} = -30\cos(4x)\sin(4x)$.

     b   Given that $y = 0$ when $x = \frac{\pi}{8}$, find an expression for y in terms of x.

3   The variables x and y are related by the differential equation $\frac{dy}{dx} = x^2\sqrt{y}$.

     When $x = 0$, $y = 2$. Find an expression for y in terms of x.

4   **a**   Using the result $\cos(A + B) \equiv \cos A \cos B - \sin A \sin B$, prove that in the case when $A = B = x$, $1 + \cos 2x \equiv 2\cos^2 x$.

     The variables x and y are related by the differential equation $\frac{dy}{dx} = 2\cos^2 x \cos^2 y$.

     b   When $x = 0$, $y = \frac{\pi}{4}$. Find an expression for y in terms of x.

5   **a**   Differentiate $\frac{x^2 - 1}{x^2 + 1}$.

     b   Hence, or otherwise, find the general solution to the differential equation $\frac{dy}{dx} = \frac{x}{\left(x^2 + 1\right)^2}$.

6 The variables x and y are related by the differential equation $\dfrac{dy}{dx} = \dfrac{4\sqrt{x}}{\cos y}$.

When, $x = 0$, $y = 0$. Find an expression for y in terms of x.

7 A particle is moving in such a way that its distance (x m) from a fixed point after t s is given by the formula $x = 0.3 \sin 2t - 0.4 \cos 2t$.

a Show that x satisfies the differential equation $\dfrac{d^2x}{dt^2} + 4x = 0$.

b Describe the motion of the particle.

8 A car is moving such that its velocity, v m s^{-1}, is proportional to its displacement, s m.

a Write down a differential equation for the motion of the car.

When $s = 30$ m, $v = 15$ m s^{-1} and $t = 3.6$ seconds.

b Find an expression for s in terms of t.

9 A car is slowing down.

The speed, v m s^{-1}, after t s satisfies the differential equation $\dfrac{dv}{dt} = -\dfrac{v^2}{100}$, $t \geqslant 0$.
The initial speed is 20 m s^{-1}.

a Find an expression for v in terms of t.

b Give a reason why the solution may not be valid as t becomes large.

10 **a** Write $\dfrac{5}{(2x+1)(x-2)}$ in partial fractions.

b Given that $\dfrac{dx}{dy} = \dfrac{(2x+1)(x-2)}{5e^y}$, and that, when $x = -3$, $y = 0$, show that $e^{-y} = 1 + \ln\left|\dfrac{2x+1}{x-2}\right|$.

11 The variables x and y are related by the differential equation $\dfrac{dy}{dx} = \dfrac{\cos x\, e^{\sin x}}{e^y}$.

When $x = \pi$, $y = 0$. Find an expression for y in terms of x.

12 The variables x and y are related by the differential equation

$\dfrac{dy}{dx} = \sin^2 x \cos^2 y + \cos^2 x \cos^2 y$.

Find the general solution to the differential equation.

13 The variables x and y are related by the differential equation

$\dfrac{dy}{dx} = 1 + \tan^2 x + \tan^2 y + \tan^2 x \tan^2 y$.

Show that $y + \sin y \cos y = 2\tan x + c$, where c is a constant.

14 The variables x and y are related by the differential equation $\dfrac{dy}{dx} = \dfrac{xy}{x-3}$.

Show that $y = Ae^x(x-3)^3$, where A is a constant.

15 The variables x and y are related by the differential equation $\dfrac{dy}{dx} + 4y = 8$.

Given that $y = 3$ when $x = 0$, find an expression for y in terms of x.

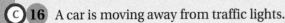

C **16** A car is moving away from traffic lights.

After t seconds, the speed of the car is v m s^{-1} and the acceleration is proportional to the difference between the speed and $20\,\text{m}\,\text{s}^{-1}$.

  a Show that $\dfrac{\mathrm{d}v}{\mathrm{d}t} = k(20 - v)$, where k is a constant.

  b Find an expression for v in terms of t.

  c What does this model imply about the speed as time increases?

PS **17** Given that $\dfrac{\mathrm{d}y}{\mathrm{d}x} = xe^{x+y}$, prove that $xe^x - e^x + \dfrac{1}{e^y}$ is a constant.

18 Let $f(x, y) = x^2 - x^2\,e^{-y} - 1 + e^{-y}$.

  a Write $f(x, y)$ as the product of two non-linear factors.

  b Hence, or otherwise, find the general solution to the differential equation $\dfrac{\mathrm{d}y}{\mathrm{d}x} = f(x, y)$.
    Give your answer in the form $y = g(x)$.

19 **a** Write $\dfrac{x}{y^2} - xy - \dfrac{1}{y^2} + y$ in the form $f(x)g(y)$.

  Let $\dfrac{\mathrm{d}y}{\mathrm{d}x} = \dfrac{x}{y^2} - xy - \dfrac{1}{y^2} + y$.

  b Show that $\ln\left|1 - y^3\right| = -\dfrac{3x^2}{2} + 3x + c$, where c is a constant.

20 By writing $\dfrac{x}{2y-1} - \dfrac{2x}{3y-2}$ in the form $f(x)g(y)$, find the general solution to the differential

  equation $\dfrac{\mathrm{d}y}{\mathrm{d}x} = \dfrac{x}{2y-1} - \dfrac{2x}{3y-2}$.

Mathematics in life and work

The insulation of a building is being tested. The temperature inside the building after t minutes is $y\,°C$ and it is initially $20\,°C$.

1 Show that the temperature satisfies the differential equation $\dfrac{\mathrm{d}y}{\mathrm{d}t} = -c\,(y - r),\ c > 0$,
  where r is the outside temperature, explaining clearly any signs.

2 Show that $y = 10e^{-ct} + r$ is a solution to the differential equation.

3 The external temperature is $10\,°C$. After 5 minutes, the temperature is $15\,°C$. Work
  out the values of c and r. What is the temperature of the room after 20 minutes?

9 COMPLEX NUMBERS

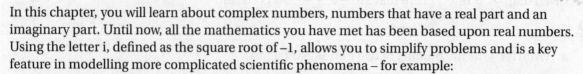

Mathematics in life and work

In this chapter, you will learn about complex numbers, numbers that have a real part and an imaginary part. Until now, all the mathematics you have met has been based upon real numbers. Using the letter i, defined as the square root of −1, allows you to simplify problems and is a key feature in modelling more complicated scientific phenomena – for example:

> If you were a physicist, you could model the relationship between electric and magnetic fields to investigate the electromagnetic spectrum, such as infrared, ultraviolet, microwaves and radio waves.

> If you were a meteorologist, you could use complex numbers to model chaotic systems, where microscopic changes in initial conditions may result in macroscopic changes in the future – this is known as 'the butterfly effect'.

> If you were a research scientist, you could use complex numbers to solve second-order differential equations resulting from real-world situations, such as simple harmonic motion and damping in springs.

In this chapter, you will consider the applications of complex numbers to simple harmonic motion and electronics.

LEARNING OBJECTIVES

You will learn how to:

> understand the idea of a complex number, recall the meaning of the terms real part, imaginary part, modulus, argument, conjugate, and use the fact that two complex numbers are equal if and only if both real and imaginary parts are equal

> carry out operations of addition, subtraction, multiplication and division of two complex numbers expressed in Cartesian form $x + iy$

> use the result that, for a polynomial equation with real coefficients, any non-real roots occur in conjugate pairs

> represent complex numbers geometrically by means of an Argand diagram

> carry out the operations of multiplication and division of two complex numbers expressed in polar form $r(\cos\theta + i\sin\theta) \equiv re^{i\theta}$

> find the two square roots of a complex number

> understand in simple terms the geometrical effects of conjugating a complex number and of adding, subtracting, multiplying and dividing two complex numbers

> illustrate simple equations and inequalities involving complex numbers by means of loci in an Argand diagram.

LANGUAGE OF MATHEMATICS

Key words and phrases you will meet in this chapter:

> Argand diagram, argument, complex number, conjugate, half line, imaginary part, locus, modulus, real part, polar form

PREREQUISITE KNOWLEDGE

You should already know how to:

> factorise expressions including the difference of two squares

> solve quadratic equations by factorisation or by completing the square

> add and subtract vectors and multiply a vector by a scalar

> calculate the magnitude of a vector

> describe a translation by using a column vector

> reflect simple plane figures in horizontal lines and rotate simple plane figures about the origin through multiples of 90°

> convert between degrees and radians

> find the equation of a straight line

> use the factor theorem.

You should be able to complete the following questions correctly:

1 Expand these expressions.

   a $(3x + 2)(4x - 9)$

   b $(4x - 9)^2$

2 Solve by factorisation.

   a $49x^2 - 64 = 0$

   b $x^2 = 4(x + 15)$

   c $5x^2 + 11x = 12$

3 Given $\mathbf{p} = 2\mathbf{i} - 7\mathbf{j}$ and $\mathbf{q} = -4\mathbf{i} + 11\mathbf{j}$, find:

   a $\mathbf{p} + \mathbf{q}$

   b $\mathbf{p} - \mathbf{q}$

   c $4\mathbf{p}$.

4 Calculate the magnitude of the vector $(5\mathbf{i} + 10\mathbf{j})$ and the angle it makes with the x-axis.

5 Find the equation of the perpendicular bisector of the points $(7, 2)$ and $(11, -8)$.

   Give the answer in the form $ax + by + c = 0$, where a, b and c are integers.

PURE MATHEMATICS 3

9.1 Definition of a complex number

In the past, you would have been told that you cannot find the square root of a negative number, that $x^2 + 6x + 13 = 0$ has no real roots because $b^2 - 4ac < 0$. The important word here is 'real' – in fact, the equation does have roots but they are complex.

To solve an equation such as $x^2 = -1$, you need to extend the concept of numbers to imaginary numbers, where i is defined as $\sqrt{-1}$.

Hence i is the solution to the equation $x^2 = -1$.

Also $\sqrt{-49} = \sqrt{49}\sqrt{-1} = \pm 7i$ and $\sqrt{-16} = \sqrt{16}\sqrt{-1} = \pm 4i$.

A **complex number** has two parts, a **real part**, Re z, and an **imaginary part**, Im z. For the complex number, $3 - 8i$, Re $z = 3$ and Im $z = -8$. In general, the complex number $x + iy$ has Re $z = x$ and Im $z = y$. Note that the i is not included as part of Im z.

Note that two complex numbers are identical if and only if their real parts are identical and their imaginary parts are identical.

A complex number can be represented on an **Argand diagram**, which is a coordinate grid. The horizontal axis is the real axis and the vertical axis is the imaginary axis.

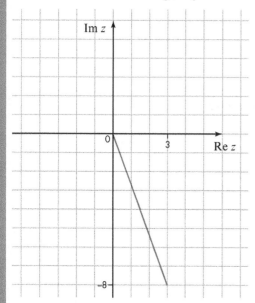

The **conjugate** of a complex number has the same real part but the opposite sign for the imaginary part. For example, $3 + 8i$ is the conjugate of $3 - 8i$. The conjugate is the mirror image of the original complex number in the real axis.

> **KEY INFORMATION**
> $i = \sqrt{-1}$ and $i^2 = -1$.

> **KEY INFORMATION**
> A complex number has a real part, Re z, and an imaginary part, Im z.

> **KEY INFORMATION**
> Two complex numbers are identical if and only if their real parts and imaginary parts are identical.

> **KEY INFORMATION**
> A complex number can be represented on an Argand diagram, using the horizontal axis for the real axis and the vertical axis for the imaginary axis.

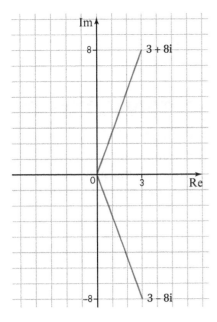

As will be discussed in **Section 9.3**, complex roots of polynomial equations come in conjugate pairs. For example, the roots of the quadratic equation $x^2 + 6x + 13 = 0$ are $-3 + 2i$ and $-3 - 2i$.

If a complex number is represented by z, its conjugate is represented by z^*. Hence if the complex number $z = x + iy$, then $z^* = x - iy$.

The **modulus** of a complex number is its magnitude. As can be seen from the Argand diagram below, the modulus can be found using Pythagoras' theorem in the same way as the magnitude of a column vector can be found.

KEY INFORMATION

The complex conjugate of $z = x + iy$ is given by $z^* = x - iy$.

KEY INFORMATION

The modulus of a complex number is its magnitude.

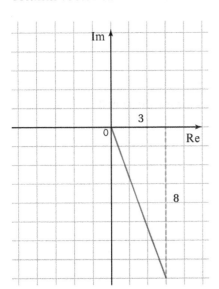

The modulus of the complex number $3 - 8i = \sqrt{3^2 + 8^2} = \sqrt{73}$.

The modulus of the complex number z is represented by $|z|$ and is given by $|z| = \sqrt{x^2 + y^2}$. Note that since it is the magnitudes of Re z and Im z that are important, you do not need to include the signs when using this formula (and x^2 and y^2 will be positive anyway).

The **argument** of a complex number is the angle made with the positive real axis and it is measured anticlockwise. Since one revolution around the Argand diagram is 2π radians, the argument takes any value within a domain of length 2π. Most often the argument will take a value in the domain $-\pi < \theta \leqslant \pi$ radians.

> **KEY INFORMATION**
>
> The argument of a complex number is the angle made with the positive real axis.

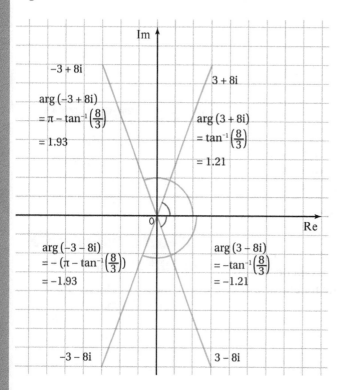

$-3 + 8i$

$\arg(-3 + 8i)$

$= \pi - \tan^{-1}\left(\dfrac{8}{3}\right)$

$= 1.93$

$3 + 8i$

$\arg(3 + 8i)$

$= \tan^{-1}\left(\dfrac{8}{3}\right)$

$= 1.21$

$\arg(-3 - 8i)$

$= -\left(\pi - \tan^{-1}\left(\dfrac{8}{3}\right)\right)$

$= -1.93$

$\arg(3 - 8i)$

$= -\tan^{-1}\left(\dfrac{8}{3}\right)$

$= -1.21$

$-3 - 8i$

$3 - 8i$

Stop and think If you were asked to write each of these arguments for the domain $0 \leqslant \theta < 2\pi$ radians instead, what would each argument be?

Example 1

Find the conjugate, modulus and argument for each of the following complex numbers.

a $z = 8 - 2i$

b $z = -5 + 12i$

c $z = 8 + 8\sqrt{3}\,i$

Solution

a The conjugate of $8 - 2i$ is $8 + 2i$.

To find the modulus, start by plotting $8 - 2i$ on an Argand diagram.

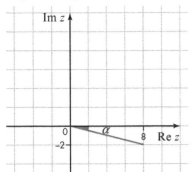

Use Pythagoras' theorem.

$|z| = \sqrt{8^2 + 2^2} = \sqrt{68} = 2\sqrt{17}$

> You can use the positive values of x and y when finding the modulus of $x + iy$.

Use trigonometry to find the argument.

From the diagram, $\alpha = \tan^{-1}\left(\dfrac{2}{8}\right) = 0.245$

If $-\pi < \arg z \leqslant \pi$, then $\arg z = -0.245$

> Remember that the argument is measured anticlockwise from the positive real axis. Since the argument is clockwise from the real axis, it is negative.

b The conjugate of $-5 + 12i$ is $-5 - 12i$.

Plot $-5 + 12i$ on an Argand diagram.

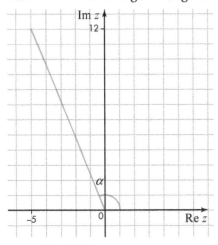

$|z| = \sqrt{5^2 + 12^2} = 13$

$\alpha = \tan^{-1}\left(\dfrac{12}{5}\right) = 1.176$

Note that the argument is obtuse.

$\arg z = \pi - 1.176 = 1.97$

> If the argument is obtuse, find the acute angle and subtract it from π.

c The conjugate of $8 + 8\sqrt{3}\,i$ is $8 - 8\sqrt{3}\,i$.

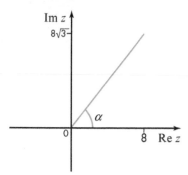

$$|z| = \sqrt{8^2 + \left(8\sqrt{3}\right)^2} = 16$$

$$\alpha = \tan^{-1}\left(\frac{8\sqrt{3}}{8}\right) = \tan^{-1}\left(\sqrt{3}\right) = \frac{\pi}{3}$$

In this example, the argument can be given in terms of π.
Recall from **Pure Mathematics 1 Chapter 4 Circular measure and trigonometry** that there are special angles that have certain values for sin, cos and tan that you should know.

Exercise 9.1A

1 Write down Re z and Im z for each complex number.

a $z = -6 + 4i$

b $z = 13 - 5i$

c $z = -8 - 9i$

d $z = 2 + 15i$

e $z = 3 - 10i$

f $z = 11 + 4i$

g $z = -16 - 2i$

h $z = -1 + 7i$

2 Plot each complex number from **Question 1** on the same Argand diagram.

3 Write down the conjugate (z^*) for each of the following complex numbers.

a $z = 4 + 2i$

b $z = 9 - i$

c $z = -15 + 3i$

d $z = 8 - 7i$

e $z = 12 + 17i$

f $z = -3 - 25i$

g $z = 20 - 20i$

h $z = -8 + \sqrt{3}\,i$

4 Find $|z|$ and arg z for each complex number.

Write $|z|$ in surd form and write the argument correct to 3 significant figures.

Your argument should be given in the domain $-\pi \leqslant \theta < \pi$.

a $z = 6 + 5i$

b $z = 10 - 7i$

c $z = -4 + 9i$

d $z = -2 - 3i$

Ⓒ Communication   ⓂⓂ Mathematical modelling   ⓅⓈ Problem solving

5 Find the modulus and the argument for each complex number.

The argument should be given in terms of π in the domain $-\pi \leqslant \theta < \pi$.

a $z = -5 + 5i$

b $7\sqrt{3} - 7i$

c $z = 11 + 11i$

d $6 - 6\sqrt{3}\,i$

6 Find each complex number in the form $x + iy$.

a $|z| = 9\sqrt{2}$, $\arg z = \dfrac{\pi}{4}$

b $|z| = 10$, $\arg z = \dfrac{\pi}{3}$

c $|z| = 6$, $\arg z = \dfrac{5\pi}{6}$

d $|z| = 4\sqrt{2}$, $\arg z = -\dfrac{3\pi}{4}$

7 For the complex number $z = -8 + 8i$, find:

a $|z^*|$

b $\arg z^*$, in the domain $-\pi \leqslant \theta < \pi$.

8 The complex number z is such that $|z| = 10\sqrt{2}$ and $\arg z = \dfrac{3\pi}{4}$.

a Write z in the form $x + iy$.

b Plot z on an Argand diagram.

c Find the size of the reflex angle between z and z^*.

9 The Argand diagram shows the complex number z_1.

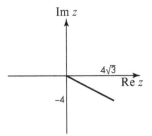

Find the complex number z_2, in the form $x + iy$, such that $|z_2| = 40$ and $\arg z_2 = \arg z_1{}^*$.

10 The complex number z is given by $-2 + 5\sqrt{2}\,i$.

a Plot z on an Argand diagram.

b Show that $|z| = 3\sqrt{6}$.

c Show that $\arg z = 1.85$, correct to 3 significant figures.

11 The complex number z_1 is such that $|z_1| = 4\sqrt{5}$ and $\arg z_1 = \tan^{-1}(2)$.

a Write z_1 in the form $x + iy$.

The complex numbers z_2 and z_3 are given by $7 - 3i$ and $-5 + i$, respectively.

A triangle is drawn with vertices at z_1, z_2 and z_3.

b Find the area of the triangle.

PURE MATHEMATICS 3

9.2 Addition, subtraction, multiplication and division of complex numbers

To add two complex numbers together, add their real parts together and add their imaginary parts together.

$z_1 + z_2 = (\text{Re } z_1 + \text{Re } z_2) + (\text{Im } z_1 + \text{Im } z_2)i$

For example, $(5 + 3i) + (6 - 9i) = (5 + 6) + (3 + (-9))i = 11 - 6i$.

The same process is used for subtraction.

$z_1 - z_2 = (\text{Re } z_1 - \text{Re } z_2) + (\text{Im } z_1 - \text{Im } z_2)i$

For example, $(5 + 3i) - (6 - 9i) = (5 - 6) + (3 - (-9))i = -1 + 12i$.

To multiply two complex numbers, you can use the same method as you have used for expanding two pairs of brackets in algebra.

For example,

$(2 + i)(5 - 2i) = 10 - 4i + 5i + 2$

$= 12 + i$.

> **KEY INFORMATION**
>
> To add or subtract complex numbers, add or subtract the real parts and the imaginary parts.

> **KEY INFORMATION**
>
> Remember that $i^2 = -1$.

This expression has two real parts and two imaginary parts, so it can be simplified.

Example 2

$z_1 = 3 + 5i$ and $z_2 = 7 - 2i$

Find simplified expressions for:

a $z_1 + z_2$

b $2z_2 - z_1{}^*$

c $z_1 z_2$

d $z_2{}^2$.

Solution

a $z_1 + z_2 = (3 + 5i) + (7 - 2i)$

$(3 + 5i) + (7 - 2i) = (3 + 7) + (5 - 2)i = 10 + 3i$

b $2z_2 - z_1{}^* = 2(7 - 2i) - (3 - 5i)$

$= (14 - 3) + (-4 + 5)i = 11 + i$

$z_1{}^*$ is the conjugate of z_1.

c $(3 + 5i)(7 - 2i) = 21 - 6i + 35i - 10i^2$

$= 21 - 6i + 35i + 10$

$= 31 + 29i$

d $(7 - 2i)^2 = (7 - 2i)(7 - 2i)$

$= 49 - 14i - 14i + 4i^2$

$= 49 - 14i - 14i - 4$

$= 45 - 28i$

Squaring a complex number is like squaring brackets.

Exercise 9.2A

1 Work out:

   a $(6 + 8i) + (3 - 5i)$ **b** $(4 - 10i) + (-1 + 3i)$

   c $(7 + 2i) - (3 - 11i)$ **d** $(7 - 8i) + (9 - 6i)$

   e $(-14 + 30i) + (12 + 17i)$ **f** $(5 - 9i) - (-4 + 7i)$

   g $(-5 - 2i) - (19 - 16i)$ **h** $(3 - 7i) + (-9 - 16i) - (-2 + 10i)$.

2 Simplify each expression, writing the answer in the form $x + iy$.

   a $(5 + 2i)(3 + i)$ **b** $(7 - 8i)(4 - 2i)$

   c $(4 - 3i)(1 + 3i)$ **d** $(10 - i)(5 + i)$

   e $(4 + 5i)^2$ **f** $(9 - 2i)^2$

   g $(2 + i)(7 - 2i)(3 + 5i)$ **h** $(4 - 6i)(1 - 2i)(2 - 3i)$

3 Simplify each expression.

   a $(5 + 7i) + (2 - 4i)$ **b** $(5 + 7i) - (2 - 4i)$ **c** $(5 + 7i)(2 - 4i)$

4 Find the value of a and b.

   a $(a + 6i) + (7 - bi) = -2 + 5i$ **b** $(a + bi) + (8 - 9i) = 10 + 3i$

   c $(11 - ai) - (b - 9i) = 7 - 6i$ **d** $(a + bi) - (-3 + 4i) = -5 + 8i$

   e $(5 - 4i) - (a + bi) = 6 - 2i$

Ⓒ 5 Show that $(7 + 2i)^3 = 259 + 286i$.

6 $z_1 = 4 - i;\ z_2 = 2 + 3i$

   Find a simplified expression for each complex number.

   a $z_1 + 3z_2$ **b** $z_2 - z_1$

   c $z_1 z_2$ **d** $z_2{}^3$

   e $z_1{}^2 - z_2{}^2$

7 The complex number z is such that $z(3 + 2i) = 28 - 3i$.

   Write z in the form $x + iy$.

⟨MM⟩ 8 $z_1 = 4 - 3i$

   Find the magnitude and argument of each complex number.

   a z_1 **b** $z_1{}^2$

   c $z_1{}^3$ **d** $z_1{}^4$

   Plot each answer in an Argand diagram.

⟨PS⟩ 9 Given that $\dfrac{a - 6i}{1 - 2i} = b + 4i$, find the values of a and b.

10 The complex number z_1 has $|z_1| = 6$ and $\arg z_1 = \frac{\pi}{4}$.

The complex number z_2 has $|z_2| = 4\sqrt{3}$ and $\arg z_2 = \frac{\pi}{6}$.

a Find $z_1 z_2$ in the form $x + iy$. **b** Show that $\arg z_1 z_2 = \tan^{-1}(2 + \sqrt{3})$.

c Verify that $\arg z_1 z_2 = \frac{5\pi}{12}$.

11 a Expand and simplify $(p - 2i)^4$ in descending powers of p.

b Given that $(p - 2i)^4 = -119 - 120i$ and that p is real, find the values of p.

Dividing a complex number by another complex number

Consider multiplying a complex number $z = x + iy$ by its conjugate $z^* = x - iy$.

$(x + iy)(x - iy) = x^2 - ixy + ixy - i^2y^2.$

$\qquad\qquad = x^2 + y^2$

Hence multiplying a complex number by its conjugate results in a real answer because the imaginary terms cancel out.

This process is used to divide complex numbers. Multiply the numerator and denominator by the conjugate of the denominator, as shown in **Example 3**.

Notice the similarity between this method for dividing complex numbers and the way that you rationalise the denominator of a surd using the difference of two squares.

> **KEY INFORMATION**
>
> Dividing complex numbers uses a similar process to rationalising the denominator of a surd, by multiplying the numerator and denominator by the conjugate of the denominator.

Example 3

Calculate $(11 + 2i) \div (2 - i)$.

Solution

Start by writing the question as $\frac{11 + 2i}{2 - i}$.

The conjugate of the denominator is $(2 + i)$.

Multiply the numerator and denominator by $(2 + i)$. ◀

> The conjugate of $(2 - i)$ is $(2 + i)$.

$\dfrac{11 + 2i}{2 - i} \times \dfrac{2 + i}{2 + i} = \dfrac{(11 + 2i)(2 + i)}{(2 - i)(2 + i)}$

> Notice that the denominator is the difference of two squares.

$\qquad = \dfrac{22 + 11i + 4i + 2i^2}{4 + 2i - 2i - i^2}$

$\qquad = \dfrac{22 + 11i + 4i - 2}{4 + 1}$

$\qquad = \dfrac{20 + 15i}{5}$

$\qquad = 4 + 3i$

Exercise 9.2B

1 Write each of the following in the form $x + iy$.

  a $(3 + 7i) \div (-1 + i)$

  b $(-5 + 14i) \div (3 + 2i)$

  c $(54 + 2i) \div (-2 - 6i)$

  d $(53 - 31i) \div (7 - 3i)$

2 Write each of the following in the form $x + iy$.

  a $\dfrac{48 + 19i}{5 - 4i}$

  b $\dfrac{17(i - 3)}{3 + 5i}$

  c $\dfrac{10}{3 - i}$

  d $\dfrac{19 - 77i}{7 - 5i}$

3 $z_1 = 12 + 5i$; $z_2 = 2 + 3i$.

  Write each answer in the form $x + iy$.

  a $z_1 + z_2$

  b $z_1 - z_2$

  c $z_1 z_2$

  d $\dfrac{z_1}{z_2}$

4 Find the value of a and b given that:

  a $(a + bi)(2 - i) = 5 + 5i$

  b $(a + bi)(1 - i) = 5 - 5i$

  c $(3 - 2i)(a + bi) = 6 - 17i$

  d $(a + bi)(1 + 3i) = -11 + 7i$.

5 Divide $(2 - i)$ by its conjugate.

6 $z_1 = 2 - 3i$; $z_2 = 21 + i$; $z_3 = 17(1 - i)$.

  Find the magnitude and argument of:

  a $z_4 = \dfrac{z_2}{z_1}$

  b $z_5 = \dfrac{1}{z_3}$

  c $z_6 = \dfrac{z_3}{z_4^*}$.

7 The complex number z_1 is such that $|z_1| = 5\sqrt{5}$ and $\arg z_1 = \tan^{-1}\left(\dfrac{1}{2}\right)$.

  Given that $\dfrac{z_1}{3 + 4i} = p + qi$, show that $p + q = 1$.

(PS) 8 Find $\left|\dfrac{1168 - 9i}{8 - 9i}\right|$.

9 The complex numbers $(u - 9i)$ and $(3 + i)$ satisfy the equation $\left|\dfrac{u - 9i}{3 + i}\right| = 5$.

  Given that $u < 0$, find the value of u.

10 The complex numbers z_1 and z_2 are given by $a + 2i$ and $5 + bi$, respectively.

  a Find $\dfrac{z_1}{z_2}$ in terms of a and b.

  b Given that $\dfrac{z_1}{z_2} = q - 2i$, show that $b^2 = \dfrac{1}{2}(ab - 60)$

  c Given further that $a = 23$ and $q > 0$, find both possible values for b.

  d Show that one possible value for q is 3 and find the other possible value for q.

PURE MATHEMATICS 3

Square roots of complex numbers

Unlike taking the square root of a negative number, taking the square root of a complex number does not require a new number system to be developed. The square root of a complex number is still a complex number. Just like real numbers, complex numbers have two square roots.

> **KEY INFORMATION**
>
> Complex numbers have two square roots, which are also complex numbers.

Example 4

Find the square roots of $21 - 20i$.

Solution

Start by writing the square root of z as $x + iy$, where x and y are real numbers.

$(x + iy)^2 = 21 - 20i$

Expand the brackets.

$x^2 + 2ixy - y^2 = 21 - 20i$

The real parts must be the same and the imaginary parts must be the same.

$\text{Re } z = x^2 - y^2 = 21$

$\text{Im } z = 2xy = -20$

Hence you have two simultaneous equations, $x^2 - y^2 = 21$ and $2xy = -20$.

> Equate the real coefficients and the imaginary coefficients.

From the second equation, $y = -\dfrac{10}{x}$.

Substitute $y = -\dfrac{10}{x}$ into the first equation.

> You could also substitute for x.

$x^2 - \left(-\dfrac{10}{x}\right)^2 = 21$

Multiply through by x^2

$\qquad x^4 - 100 = 21x^2$

$x^4 - 21x^2 - 100 = 0$

$(x^2 - 25)(x^2 + 4) = 0$

$\qquad x^2 = 25 \text{ or } -4$

> Note that $x^2 = -4$ has no real solutions.

Since x is a real number, $x = \pm 5$.

When $x = 5$, $y = -2$.

When $x = -5$, $y = 2$.

The square roots of $21 - 20i$ are $5 - 2i$ and $-5 + 2i$.

Exercise 9.2C

1 Find the square roots of each complex number.

   a $3 - 4i$ **b** $-15 + 8i$

   c $5 - 12i$ **d** $-7 - 24i$

2 Find the square roots of each complex number.

   a $85 + 132i$ **b** $336 + 320i$

3 The square roots of $55 - 48i$ are $(a - 3i)$ and $(b + ci)$.

   Find the values of a, b and c.

4 Angelene is finding the square roots of $(65 - 72i)$.

   This is her method:

$(a + bi)^2 = 65 - 72i$

   $a^2 + abi + abi + bi^2 = 65 - 72i$

   $a^2 - b^2 = 65$

   $2abi = -72i$

   $a = \dfrac{-72i}{2bi}$

   $\left(\dfrac{-72i}{2bi}\right)^2 - b^2 = 65$

   $(-72i)^2 - (2bi)^2 b^2 = 65(2bi)^2$

   $-72i \times -72i - 2bi \times 2bi \times b^2 = 65 \times 2bi \times 2bi$

   $5184(-1) - 4b^4(-1) = 260b^2(-1)$

   $5184 - 4b^4 = 260b^2$

   $5184 - 260b^2 - 4b^4 = 0$

   $-1296 + 65b^2 + b^4 = 0$

   $b^4 + 65b^2 - 1296 = 0$

   $(b^2 + 32.5)^2 - 32.5^2 - 1296 = 0$

   $(b^2 + 32.5)^2 - 1056.25 - 1296 = 0$

   $(b^2 + 32.5)^2 - 2352.25 = 0$

   $(b^2 + 32.5)^2 = 2352.25$

   $b^2 + 32.5 = \pm 48.5$

   $b^2 = \pm 48.5 + 32.5$

   $b^2 = -16$ or 81

   b^2 cannot be -16

   $b^2 = 81$

   $b = 9$ or -9

   $a = -4$ or 4

   Comment on Angelene's solution.

5 **a** Expand and simplify $(p + qi)^2$.

    b Given that $(p + qi)^2 = 45 + mi$, and that p and q are positive integers with $p > q$, find the three possible values for m.

    c Hence state the square roots of the complex number $45 - 108i$.

6 **a** Show that one of the square roots of $(7 + 24i)$ is $(4 + 3i)$ and find the other square root.

    b Show that the argument of $(7 + 24i)$ is twice the argument of $(4 + 3i)$.

    c Show that the modulus of $(7 + 24i)$ is the square of the modulus of $(4 + 3i)$.

7 **a** Find the square roots of the complex number $(105 + 88i)$.

    b Hence find the square roots of the complex number $(945 + 792i)$.

8 Given that $z = \dfrac{-262 + 130i}{7 + 5i}$, find the square roots of z.

9 Evaluate $\left| \sqrt{\dfrac{362 - 153i}{2 - 3i}} \right|$.

10 The complex number $z_1 = 40 - 42i$.

    a Find $|z_1|$.   **b** Find $\arg z_1$.

    z_2 and z_3 are the square roots of z_1 such that $\operatorname{Re} z_2 > 0$.

    c Show that $\operatorname{Re} z_2 = 7$ and find $\operatorname{Im} z_2$.   **d** Hence find $|z_3|$ and $\arg z_3$.

9.3 Complex roots of polynomial equations

In your previous study you solved quadratic equations. When $b^2 - 4ac > 0$ there are two real and distinct roots; when $b^2 - 4ac = 0$ there are two real equal roots; and when $b^2 - 4ac < 0$ there are no real roots. Even though there are no real roots, there are roots – they just happen to be imaginary or complex.

All quadratic equations have two roots, whether they are real, equal or complex. In fact, all polynomial equations have the same number of solutions as the highest degree of the terms. A cubic equation has three roots, a quartic (x^4) equation has four roots, and so on.

Complex roots come in pairs: a complex number and its conjugate (see **Section 9.1**). Hence if you know one complex root, you know another. For example, if one root of a quadratic equation is $2 - 9i$, the other is $2 + 9i$.

A cubic equation has either three real roots or one real root and a pair of complex roots.

A quartic equation has either four real roots or two real roots and a pair of complex roots or two pairs of complex roots.

KEY INFORMATION

Complex roots come in pairs, a complex number and its conjugate.

Stop and think   What could you say about the roots of the equation $x^6 - x^4 - 16x^3 - 17x^2 - 16x - 15 = 0$?

Example 5

Find all the roots of the equation $z^3 + 4z^2 + z = 26$, given that one root is 2.

Solution

This is a cubic equation, so it has three roots. You are told that one root is 2. The other roots are either also real or they are a pair of imaginary or complex numbers.

Write the equation as $z^3 + 4z^2 + z - 26 = 0$.

Since one root is 2, $(z - 2)$ is a factor of $z^3 + 4z^2 + z - 26 = 0$.

By long division:

$$
\begin{array}{r}
z^2 + 6z + 13 \\
z - 2 \overline{\smash{\big)}\ z^3 + 4z^2 + z - 26} \\
\underline{z^3 - 2z^2} \\
6z^2 + z \\
\underline{6z^2 - 12z} \\
13z - 26 \\
\underline{13z - 26} \\
0
\end{array}
$$

Hence $z^3 + 4z^2 + z - 26 \equiv (z - 2)(z^2 + 6z + 13)$.

$z^3 + 4z^2 + z - 26 = 0$

$(z - 2)(z^2 + 6z + 13) = 0$

Either $z - 2 = 0$ or $z^2 + 6z + 13 = 0$.

Solve the quadratic equation by completing the square.

$(z + 3)^2 - 9 + 13 = 0$

$(z + 3)^2 + 4 = 0$

$(z + 3)^2 = -4$

Take the square root.

$z + 3 = \pm 2i$

$z = -3 \pm 2i$

The roots of the equation are 2 and $-3 \pm 2i$.

> You could also find the quadratic factor by comparing terms. The first term must be z^2 so that the cubic begins with z^3 and the last term must be $+13$ so that the cubic ends with -26. The middle term can then be found by comparing coefficients.

> Note that $b^2 - 4ac = 6^2 - 4 \times 1 \times 13 = -16$. Since $b^2 - 4ac < 0$, the roots are complex.

> The quadratic equation can also be solved using the quadratic formula, but you would not be able to factorise it.

Example 6

The quadratic equation $x^2 + ax + b = 0$ has a root $3 + 9i$. Find the values of a and b.

Solution

Firstly, note that if one root is $3 + 9i$, then another root is given by the complex conjugate, $3 - 9i$.

You can find the quadratic equation by reversing the process.

Start with $x = 3 \pm 9i$.

$x - 3 = \pm 9i$

$(x - 3)^2 = -81$

$x^2 - 6x + 9 = -81$

$x^2 - 6x + 90 = 0$

The quadratic equation $x^2 - 6x + 90 = 0$ has the roots $3 \pm 9i$.
Hence $a = -6$ and $b = 90$.

Here is an alternative method to find the quadratic equation.
$(x - (3 + 9i))(x - (3 - 9i)) = 0$
This can be rewritten as the difference of 2 squares.
$((x - 3) + 9i)((x - 3) - 9i) = 0$
$(x - 3)^2 - (9i)^2 = 0$
$x^2 - 6x + 90 = 0$

Exercise 9.3A

1 One root of the equation $z^2 - 8z + k = 0$ is $(4 + 11i)$.

   a Write down the other root.

   b Find the value of k.

c 2 Alan is trying to solve the quadratic equation $z^2 - 14z + 58 = 0$.

   a Show that $b^2 - 4ac < 0$.

   b Find the roots of the quadratic equation.

3 Solve the following equations.

   a $z^2 = -25$ **b** $z^2 + 144 = 0$

   c $z^2 + 10z + 26 = 0$ **d** $z^2 - 14z + 53 = 0$

   e $z^2 + 8z + 80 = 0$ **f** $z^2 + 12z + 37 = 0$

   g $z^2 + 104 = 20z$ **h** $z^2 + 18z + 202 = 0$

   i $z^2 + 41 = 10z$ **j** $z^2 - 12z + 4936 = 0$

   k $2z^2 + 17 = 10z$ **l** $9z^2 + 68 = 48z$

4 Find quadratic equations that have the following roots.

   a $2 \pm 5i$ **b** $7 \pm 4i$

   c $-8 \pm 20i$ **d** $-3 \pm 2i$

5 **a** Solve the equation $x^3 + x^2 + 15x = 225$, given that one root is 5.

   b Solve the equation $x^3 + 7x^2 - 13x + 45 = 0$, given that one root is -9.

   c Solve the equation $x^3 + 10x^2 + 29x + 30 = 0$, given that one root is $-2 + i$.

   d Solve the equation $x^3 - 10x^2 + 58x - 136 = 0$, given that one root is $3 - 5i$.

   e Solve the equation $3x(x^2 + 45) = 2(19x^2 + 37)$, given that one root is $6 - i$.

   f Solve the equation $x^4 + 3x^3 + x^2 + 13x + 30 = 0$, given that one root is $1 + 2i$.

6 Solve each equation.

a $z^3 - 8z^2 + 9z - 72 = 0$ **b** $z^3 + 4z + 10 = 5z^2$ **c** $2z^3 = 8z^2 + 13z - 87$

 7 **a** **i** Solve the equation $z^3 + 60 = z(5z + 4)$.

      ii Plot the roots on an Argand diagram.

    b **i** Solve the quartic equation $z^4 + z^2 + 2z = 2z^3 + 2$.

      ii Plot the roots on an Argand diagram.

8 **a** Given that $x = \frac{2}{3}$ is a root of the equation $3x^3 + 10x^2 + 16x = c$, find the value of c and solve the equation.

    b Given that $z = 1 + 7i$ is a root of the equation $2z^3 + kz^2 + 102z = 50$, find the value of k and solve the equation.

9 The complex number $z_1 = \dfrac{49 - 59i}{2 - 13i}$. z_1 is a root of the equation $z^3 + 94z = 16z^2 + 204$.
Find the real root of the equation.

10 z_1 is the square root of the complex number $-27 - 36i$ for which $\operatorname{Im} z_1 > 0$.

    a Find z_1, giving the answer in the form $x + iy$.

    b Given that z_1 is a root of the equation $z^4 + 59z^2 + 3330 = 2z(2z^2 + 3)$, find the other three complex roots of the equation.

PS **11** The complex number $z_1 = (10 - i) - (2 - 7i)$.

    a Write z_1 in the form $x + iy$.

    z_1 is one of the complex roots of the cubic equation $z^3 - 20z^2 + 164z - 400 = 0$.

    b Solve the equation $z^3 - 20z^2 + 164z - 400 = 0$.

    c Hence solve the equation $x^6 + 164x^2 = 20(x^4 + 20)$.

9.4 Polar form

As well as being written in Cartesian form, complex numbers can also be written in **polar form**, in terms of the modulus (r) and argument (θ).

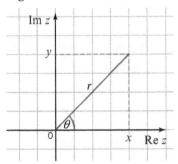

From the diagram, $x = r\cos\theta$ and $y = r\sin\theta$.

Hence $z = x + iy = r\cos\theta + ir\sin\theta = r(\cos\theta + i\sin\theta)$.

KEY INFORMATION

$z = x + iy$ can be written in polar form as $r(\cos\theta + i\sin\theta)$ and exponential form as $re^{i\theta}$.

Although the derivation is not within the scope of this course, $\cos\theta + i\sin\theta$ can be written as $e^{i\theta}$, so a complex number can also be written as $z = re^{i\theta}$.

Stop and think

How are the modulus and argument of a complex number relevant to the electric and magnetic components of a field?

Example 7

Write the following in the form $r(\cos\theta + i\sin\theta)$ and $re^{i\theta}$.

a $z = 8 + 8\sqrt{3}\,i$ **b** $z = 8 - 2i$

Solution

Note that the modulus and argument for each of these complex numbers were found in **Example 1** in **Section 9.1**.

a For $z = 8 + 8\sqrt{3}i$, $|z| = 16$ and $\arg z = \dfrac{\pi}{3}$.

Hence, $8 + 8\sqrt{3}i = 16\left(\cos\dfrac{\pi}{3} + i\sin\dfrac{\pi}{3}\right)$ or $16e^{\frac{\pi}{3}i}$.

b For $z = 8 - 2i$, $|z| = 2\sqrt{17}$ and $\arg z = -0.245$.

Hence, $8 - 2i = 2\sqrt{17}\ (\cos(-0.245) + i\sin(-0.245))$.

Note that since $\cos(-A) = \cos(A)$ and $\sin(-A) = -\sin(A)$, this can be rewritten as $2\sqrt{17}\ (\cos(0.245) - i\sin(0.245))$ or $2\sqrt{17}\,e^{-0.245i}$.

Multiplying complex numbers in polar form

Let $z_1 = r_1(\cos\theta_1 + i\sin\theta_1)$ and $z_2 = r_2(\cos\theta_2 + i\sin\theta_2)$.

$$
\begin{aligned}
z_1 z_2 &= r_1(\cos\theta_1 + i\sin\theta_1) \times r_2(\cos\theta_2 + i\sin\theta_2) \\
&= r_1 r_2(\cos\theta_1 + i\sin\theta_1)(\cos\theta_2 + i\sin\theta_2) \\
&= r_1 r_2(\cos\theta_1\cos\theta_2 + i\cos\theta_1\sin\theta_2 + i\sin\theta_1\cos\theta_2 + i^2\sin\theta_1\sin\theta_2) \\
&= r_1 r_2(\cos\theta_1\cos\theta_2 - \sin\theta_1\sin\theta_2 + i(\sin\theta_1\cos\theta_2 + \cos\theta_1\sin\theta_2)) \\
&= r_1 r_2(\cos(\theta_1 + \theta_2) + i(\sin(\theta_1 + \theta_2))
\end{aligned}
$$

This proof uses the compound angle formulae from **Chapter 3 Trigonometry**.

Note that you can prove the same result using the exponential form.

Let $z_1 = r_1 e^{i\theta_1}$ and $z_2 = r_2 e^{i\theta_2}$.

$$
\begin{aligned}
z_1 z_2 &= r_1 e^{i\theta_1} \times r_2 e^{i\theta_2} \\
&= r_1 r_2 e^{i\theta_1} \times e^{i\theta_2} \\
&= r_1 r_2 e^{i\theta_1 + i\theta_2} \\
&= r_1 r_2 e^{i(\theta_1 + \theta_2)}
\end{aligned}
$$

KEY INFORMATION

To multiply two complex numbers written in polar or exponential form, multiply the moduli and add the arguments.

Hence, to multiply two complex numbers written in polar or exponential form, you multiply the moduli and add the arguments.

Dividing complex numbers in polar form

Let $z_1 = r_1(\cos\theta_1 + i\sin\theta_1)$ and $z_2 = r_2(\cos\theta_2 + i\sin\theta_2)$.

$$\frac{z_1}{z_2} = \frac{r_1(\cos\theta_1 + i\sin\theta_1)}{r_2(\cos\theta_2 + i\sin\theta_2)}$$

$$= \frac{r_1(\cos\theta_1 + i\sin\theta_1)}{r_2(\cos\theta_2 + i\sin\theta_2)} \times \frac{(\cos\theta_2 - i\sin\theta_2)}{(\cos\theta_2 - i\sin\theta_2)}$$

$$= \frac{r_1(\cos\theta_1 + i\sin\theta_1)(\cos\theta_2 - i\sin\theta_2)}{r_2(\cos\theta_2 + i\sin\theta_2)(\cos\theta_2 - i\sin\theta_2)}$$

$$= \frac{r_1(\cos\theta_1\cos\theta_2 - i\cos\theta_1\sin\theta_2 + i\sin\theta_1\cos\theta_2 - i^2\sin\theta_1\sin\theta_2)}{r_2(\cos^2\theta_2 - i\cos\theta_2\sin\theta_2 + i\sin\theta_2\cos\theta_2 - i^2\sin^2\theta_2)}$$

$$= \frac{r_1(\cos\theta_1\cos\theta_2 + \sin\theta_1\sin\theta_2 + i(\sin\theta_1\cos\theta_2 - \cos\theta_1\sin\theta_2))}{r_2(\cos^2\theta_2 + \sin^2\theta_2)}$$

$$= \frac{r_1}{r_2}(\cos(\theta_1 - \theta_2) + i(\sin(\theta_1 - \theta_2))$$

> This proof also uses the compound angle formulae from **Chapter 3 Trigonometry**.

Hence, to divide two complex numbers written in polar or exponential form, you divide the moduli and subtract the arguments.

In summary:

$$|z_1 z_2| = |z_1||z_2|$$

$$\arg z_1 z_2 = \arg z_1 + \arg z_2$$

$$\left|\frac{z_1}{z_2}\right| = \frac{|z_1|}{|z_2|}$$

$$\arg\frac{z_1}{z_2} = \arg z_1 - \arg z_2$$

KEY INFORMATION

To divide two complex numbers written in polar or exponential form, divide the moduli and subtract the arguments.

Example 8

Given that $z_1 = 10\left(\cos\frac{2\pi}{7} + i\sin\frac{2\pi}{7}\right)$ and

$z_2 = 2\left(\cos\frac{9\pi}{14} + i\sin\frac{9\pi}{14}\right)$, find:

a $z_1 z_2$ **b** $\frac{z_1}{z_2}$.

Solution

a The moduli are 10 and 2, so $|z_1 z_2| = 10 \times 2 = 20$.

The arguments are $\frac{2\pi}{7}$ and $\frac{9\pi}{14}$, so

$$\arg z_1 z_2 = \frac{2\pi}{7} + \frac{9\pi}{14} = \frac{4\pi}{14} + \frac{9\pi}{14} = \frac{13\pi}{14}.$$

Hence $z_1 z_2 = 20\left(\cos\frac{13\pi}{14} + i\sin\frac{13\pi}{14}\right).$

b $\left|\dfrac{z_1}{z_2}\right| = 10 \div 2 = 5.$

$\arg\dfrac{z_1}{z_2} = \dfrac{2\pi}{7} - \dfrac{9\pi}{14} = \dfrac{4\pi}{14} - \dfrac{9\pi}{14} = -\dfrac{5\pi}{14}$

Hence $\dfrac{z_1}{z_2} = 5\left(\cos\left(-\dfrac{5\pi}{14}\right) + i\sin\left(-\dfrac{5\pi}{14}\right)\right).$

Since the argument is negative, this can be simplified to

$\dfrac{z_1}{z_2} = 5\left(\cos\dfrac{5\pi}{14} - i\sin\dfrac{5\pi}{14}\right).$

Exercise 9.4A

1 Write the following in both polar and exponential form.

a $-5 + 5i$

b $2\sqrt{3} - 2i$

c $7 + 3i$

d $-6 - 10i$

2 Simplify the following, giving arguments in the domain $-\pi < \theta \leqslant \pi$.

a $3(\cos 2 + i\sin 2) \times 7(\cos 3 + i\sin 3)$

b $12\left(\cos\dfrac{3}{4} + i\sin\dfrac{3}{4}\right) \times 4\left(\cos\dfrac{3}{2} - i\sin\dfrac{3}{2}\right)$

c $5e^{\frac{\pi}{3}i} \times 6e^{\frac{\pi}{5}i}$

d $\left(\cos\dfrac{3\pi}{10} + i\sin\dfrac{3\pi}{10}\right) \times \left(\cos\dfrac{2\pi}{5} + i\sin\dfrac{2\pi}{5}\right)$

e $11\left(\cos\dfrac{\pi}{6} - i\sin\dfrac{\pi}{6}\right) \times 2\left(\cos\dfrac{\pi}{2} + i\sin\dfrac{\pi}{2}\right)$

f $4(\cos(0.573) + i\sin(0.573)) \times 5(\cos(0.228) + i\sin(0.228))$

3 Simplify the following, giving arguments in the domain $-\pi < \theta \leqslant \pi$.

a $3(\cos 2 + i\sin 2) \div 7(\cos 3 + i\sin 3)$

b $12\left(\cos\dfrac{3}{4} + i\sin\dfrac{3}{4}\right) \div 4\left(\cos\dfrac{3}{2} - i\sin\dfrac{3}{2}\right)$

c $5e^{\frac{\pi}{3}i} \div 6e^{\frac{\pi}{5}i}$

d $\left(\cos\dfrac{3\pi}{10} + i\sin\dfrac{3\pi}{10}\right) \div \left(\cos\dfrac{2\pi}{5} + i\sin\dfrac{2\pi}{5}\right)$

e $11\left(\cos\dfrac{\pi}{6} - i\sin\dfrac{\pi}{6}\right) \div 2\left(\cos\dfrac{\pi}{2} + i\sin\dfrac{\pi}{2}\right)$

f $4(\cos(0.573) + i\sin(0.573)) \div 5(\cos(0.228) + i\sin(0.228))$

4 Write the following in the form $x + iy$.

a $9\left(\cos\frac{\pi}{4} - i\sin\frac{\pi}{4}\right)$

b $2\sqrt{3}e^{\frac{5\pi}{6}i}$

c $3e^{\frac{\pi}{2}i}$

d $14e^{-\frac{\pi}{3}i}$

5 $z_1 = 18\left(\cos\frac{2\pi}{3} + i\sin\frac{2\pi}{3}\right)$. $z_2 = 3\left(\cos\frac{5\pi}{12} - i\sin\frac{5\pi}{12}\right)$.

Find:

a $z_1 z_2$

b $\frac{z_1}{z_2}$.

Give the arguments in the domain $-\pi < \theta \leqslant \pi$.

c Prove the rule for dividing complex numbers when they are written in exponential form.

(c) 6 **a** Multiply $8\left(\cos\frac{\pi}{4} - i\sin\frac{\pi}{4}\right)$ by its conjugate.

b Show that $zz^* = |zz^*|$ for any complex number z.

(c) 7 $z_1 = 18\left(\cos\frac{2\pi}{3} + i\sin\frac{2\pi}{3}\right)$. $z_2 = 3\left(\cos\frac{5\pi}{12} - i\sin\frac{5\pi}{12}\right)$. Giving the arguments in the domain

$-\pi < \theta \leqslant \pi$:

a find $z_1 z_2$ **b** find $\frac{z_1}{z_2}$

c find z_1^2 **d** show that z_1^3 is real.

e Given that z_2^n is real, find the smallest possible positive integer value for n.

(c) 8 $z_1 = 3 + 3\sqrt{3}\,i$; $z_2 = 4 + 4i$.

a Find $|z_1|$ and $|z_2|$.

b Find $\arg z_1$ and $\arg z_2$.

c Write z_1 and z_2 in polar form.

d **i** Find $(3 + 3\sqrt{3}i)(4 + 4i)$ and write the answer in polar form.

   ii Find $z_1 z_2$ by multiplying the complex numbers in polar form and verify that the answer is the same as for **part d i**.

e **i** Find $\frac{3+3\sqrt{3}i}{4+4i}$ and write the answer in polar form.

   ii Find $\frac{z_1}{z_2}$ by dividing the complex numbers in polar form and verify that the answer is the same as for **part e i**.

9 $z = 8\left(\cos\frac{2\pi}{3} - i\sin\frac{2\pi}{3}\right)$.

a Plot z on an Argand diagram.

b Find the square roots of z.

10 The complex numbers z_1 and z_2 are given by $12\left(\cos\dfrac{5\pi}{6} - i\sin\dfrac{5\pi}{6}\right)$ and $4(\cos\beta - i\sin\beta)$, respectively.

Given that Re $z_2 = 2\sqrt{2}$ and $0 \leqslant \beta < \dfrac{\pi}{2}$,

a find $|z_1 z_2|$

b find $\arg\left(\dfrac{z_1}{z_2}\right)$

c plot $\left(\dfrac{z_1}{z_2}\right)$ on an Argand diagram.

Mathematics in life and work: Group discussion

You are a research scientist investigating circular motion. At time t, the position of a point is given by $z = x + iy = A\cos\omega t + iA\sin\omega t$. To find the velocity you need to differentiate.

1 Find an expression for $\dfrac{dz}{dt}$.

2 Show that $\dfrac{dz}{dt}$ is perpendicular to z.

The acceleration, a, is given by $\dfrac{d^2z}{dt^2}$.

3 Show that $a = -\omega^2 A e^{i\omega t}$.

9.5 Geometric effects

In **Section 9.2**, you were shown how to add, subtract, multiply and divide complex numbers. Now you will investigate what effect each process has on complex numbers on the Argand diagram.

Adding and subtracting complex numbers

Adding and subtracting complex numbers is similar to adding and subtracting vectors (written either as column vectors or using **i j** notation). Whereas with a column vector **i** and **j** refer to the x- and y-axes, the real and imaginary parts of a complex number refer to the real and imaginary axes. As a result, adding and subtracting complex numbers has the same effect geometrically as adding vectors, which is that the complex number is translated.

> **KEY INFORMATION**
>
> Adding and subtracting complex numbers has the effect of translating a complex number.

For example, for addition of complex numbers:
$(3 + 2i) + (5 - 9i) = 8 - 7i$.

For addition of vectors: $\begin{pmatrix} 3 \\ 2 \end{pmatrix} + \begin{pmatrix} 5 \\ -9 \end{pmatrix} = \begin{pmatrix} 8 \\ -7 \end{pmatrix}$.

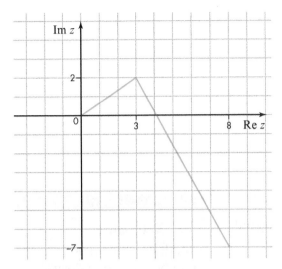

Similarly, for subtraction of complex numbers:
$(3 + 2i) - (5 - 9i) = -2 + 11i$.

For subtraction of vectors: $\begin{pmatrix} 3 \\ 2 \end{pmatrix} - \begin{pmatrix} 5 \\ -9 \end{pmatrix} = \begin{pmatrix} -2 \\ 11 \end{pmatrix}$.

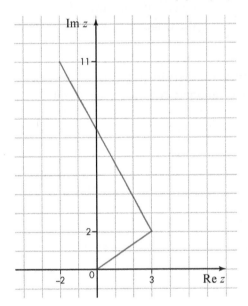

Multiplying complex numbers

You have seen in **Section 9.4** that to multiply two complex numbers in polar form you multiply the moduli and add the arguments. Again, this can be shown geometrically on the Argand diagram.

For example, $3\left(\cos \frac{\pi}{4} + i \sin \frac{\pi}{4}\right) \times 2\left(\cos \frac{\pi}{3} + i \sin \frac{\pi}{3}\right)$

$= 6\left(\cos \frac{7\pi}{12} + i \sin \frac{7\pi}{12}\right)$.

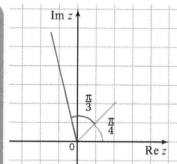

Notice that multiplying a complex number by $2(\cos\frac{\pi}{3} + i\sin\frac{\pi}{3})$:

» doubles the modulus of the complex number, as on an Argand diagram it is twice as long

» rotates the complex number by $\frac{\pi}{3}$ in an anticlockwise direction about the origin.

In general, multiplying a complex number by $a(\cos b + i\sin b)$:

» multiplies the modulus of the complex number by a

» rotates the complex number by b in an anticlockwise direction about the origin.

KEY INFORMATION

Multiplying a complex number by $a(\cos b + i\sin b)$ multiplies the modulus of the complex number by a and rotates the complex number by b in an anticlockwise direction about the origin.

Stop and think
How can you use the multiplication of complex numbers in polar form to explain why the product of a complex number and its conjugate is always a real number?

Multiplying by i

Consider multiplying by i.

$(3 + 5i) \times i = -5 + 3i$

$(-5 + 3i) \times i = -3 - 5i$

$(-3 - 5i) \times i = 5 - 3i$

$(5 - 3i) \times i = 3 + 5i$

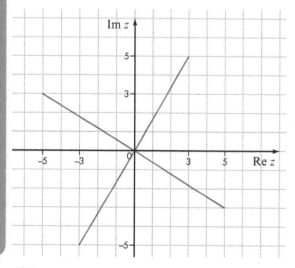

Each time a complex number is multiplied by i, the modulus is unchanged but it is rotated by $\frac{\pi}{2}$ anticlockwise about the origin.

Note that i can be written as $1(\cos\frac{\pi}{2} + i\sin\frac{\pi}{2})$.

Dividing complex numbers

To divide complex numbers in polar form, divide the moduli and subtract the arguments.

For example, $8(\cos\frac{\pi}{2} + i\sin\frac{\pi}{2}) \div 2(\cos\frac{\pi}{3} + i\sin\frac{\pi}{3}) = 4(\cos\frac{\pi}{6} + i\sin\frac{\pi}{6})$.

Notice that dividing by $2(\cos\frac{\pi}{3} + i\sin\frac{\pi}{3})$:

> halves the modulus of the complex number, so on an Argand diagram it is half as long

> rotates the complex number by $\frac{\pi}{3}$ in a clockwise direction about the origin.

In general, dividing a complex number by $a(\cos b + i\sin b)$:

> divides the modulus of the complex number by a

> rotates the complex number by b in a clockwise direction about the origin.

Dividing by i

Consider dividing by i.

$$(3 + 5i) \div i = \frac{3 + 5i}{i} = \frac{3}{i} + 5 = \frac{3i}{i^2} + 5 = \frac{3i}{-1} + 5 = 5 - 3i$$

$$(5 - 3i) \div i = \frac{5 - 3i}{i} = \frac{5}{i} - 3 = \frac{5i}{i^2} - 3 = \frac{5i}{-1} - 3 = -3 - 5i$$

and so on.

Similarly, each time a complex number is divided by i, the modulus is unchanged but it is rotated by $\frac{\pi}{2}$ clockwise about the origin.

KEY INFORMATION

Multiplying by i rotates a complex number by $\frac{\pi}{2}$ radians anti-clockwise about the origin.

KEY INFORMATION

Dividing a complex number by $a(\cos b + i\sin b)$ divides the modulus of the complex number by a and rotates the complex number by b in a clockwise direction about the origin.

KEY INFORMATION

Dividing by i rotates a complex number by $\frac{\pi}{2}$ radians clockwise about the origin.

Example 9

Consider the complex number $z = 4\sqrt{3} + 4i$.

Describe the geometric effect on another complex number of:

a adding the complex number $4\sqrt{3} + 4i$

b subtracting the complex number $4\sqrt{3} + 4i$

c multiplying by the complex number $z = 4\sqrt{3} + 4i$

d dividing by the complex number $z = 4\sqrt{3} + 4i$.

PURE MATHEMATICS 3

Solution

a Adding z has the effect of translating a complex number by $(4\sqrt{3} + 4\mathrm{i})$.

b Subtracting z has the effect of translating a complex number by $(-4\sqrt{3} - 4\mathrm{i})$.

c

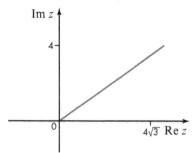

Writing z in polar form:

$$|z| = \sqrt{(4\sqrt{3})^2 + 4^2} = 8$$

$$\arg z = \tan^{-1}\left(\frac{4}{4\sqrt{3}}\right) = \tan^{-1}\left(\frac{1}{\sqrt{3}}\right) = \frac{\pi}{6}$$

$$z = 8(\cos\tfrac{\pi}{6} + \mathrm{i}\sin\tfrac{\pi}{6}).$$

Multiplying by z has the effect of multiplying the modulus by 8 and rotating anticlockwise by $\frac{\pi}{6}$.

d Dividing by z has the effect of dividing the modulus by 8 and rotating clockwise by $\frac{\pi}{6}$.

Exercise 9.5A

 1 **a** Describe the transformation that results from multiplying by each of the following.

i i **ii** i^2 **iii** i^3

iv i^4 **v** $-\mathrm{i}$ **vi** $\frac{1}{\mathrm{i}}$

vii i^{20} **viii** i^{99} **ix** $2\mathrm{i}^5$

x $0.6\mathrm{i}^{350}$

b Describe the transformation that results from dividing by each of the following.

i i **ii** i^8 **iii** $3\mathrm{i}^4$ **iv** $\frac{1}{\mathrm{i}^3}$

C 2 $z_1 = -4 + 2i$

For each operation, find the complex number and describe the geometrical effect on z_1.

a $z_1 + (2 - 3i)$ 　　　　　　　　　　　**b** $z_1{}^*$

c $z_1 \times i$ 　　　　　　　　　　　　**d** $z_1 - (2 - 3i)$

e $z_1 \div i$

C 3 $z_1 = 5\left(\cos\frac{\pi}{4} + i\sin\frac{\pi}{4}\right)$

For each operation, find the complex number and describe the geometrical effect on z_1. Each argument should be in terms of π in the domain $-\pi < \theta \leqslant \pi$.

a $z_1 \times 3\left(\cos\frac{\pi}{5} + i\sin\frac{\pi}{5}\right)$

b $z_1 \div 4(\cos 2.4 + i\sin 2.4)$

c $z_1 \times 7\left(\cos\frac{\pi}{12} - i\sin\frac{\pi}{12}\right)$

d $z_1 \div 5\left(\cos\frac{3\pi}{4} + i\sin\frac{3\pi}{4}\right)$

e $z_1{}^*$

4 z_1 is multiplied by z_2.

Write down the imaginary or complex number for z_2 that:

a rotates z_1 by 120° clockwise about the origin

b rotates z_1 by 180° about the origin

c multiplies the modulus by 5 and rotates z_1 by 450° anti-clockwise about the origin

d rotates z_1 by 90° anti-clockwise about the origin.

PS 5 $z = -2 - 2i$

Find:

a i $\arg z$ 　　　　　　　　　　　　**b i** $\arg z^5$

　　ii $|z|$ 　　　　　　　　　　　　　　**ii** $\left|z^5\right|$.

PS 6 $z = -3 + 10i$

Find:

a $\arg z$ 　　　**b** $\arg z^2$ 　　　**c** $\arg\sqrt{z}$ 　　　**d** $\arg\sqrt[3]{z}$.

7 The complex number z_1 is such that $|z_1| = 6$ and $\arg(z_1) = \frac{2\pi}{3}$.

a Write z_1 in the form $x + iy$.

z_2 is equal to iz_1.

b State the geometric relationship between z_1 and z_2.

c Find the argument of z_2 in the domain $-\pi \leqslant \arg(z_2) < \pi$.

d Prove that $|z_1 - z_2| = 6\sqrt{2}$.

8 $z_1 = -2 + 5i$

(C) $z_2 = \dfrac{-20 + 21i}{z_1}$

a Write z_2 in the form $x + iy$.

b Plot z_1 and z_2 on an Argand diagram.

z_1 and z_2 are represented by the points A and B, respectively.

c Show that OAB is a right-angled triangle.

d Find the centre of the circle that passes through O, A and B.

(C) **9** Complex numbers z_1, z_2 and z_3 are given by $z_1 = -1 + 2i$, $z_2 = 10 + 7i$ and $z_3 = 5 - 4i$.

a Plot on an Argand diagram, the points A, B and C representing z_1, z_2 and z_3, respectively.

b Show that ABC is an isosceles triangle.

10 The complex number z is given by $8\sqrt{3} - 8i$.

a Find $\arg z^2$.

b Find $\arg z^{33}$.

c Find the value of $|z^{4.5}|$.

Mathematics in life and work: Group discussion

You are a research scientist modelling the relationship between the extension (x) of a spring and time (t). The relationship is given by the second-order differential equation $m\dfrac{d^2x}{dt^2} + kx = 0$, where m is the mass of the object extending the string.

1 Show that $x = e^{\frac{3}{2}it}$ is a solution of the equation $4\dfrac{d^2x}{dt^2} + 9x = 0$.

2 What other exponential solutions does this equation have?

3 What trigonometric solutions does this equation have?

4 What general solutions would the equation $m\dfrac{d^2x}{dt^2} + kx = 0$ have?

9.6 Loci

Circles

Consider the equation $|z| = 5$. This represents all the complex numbers that have a modulus of 5. If these were plotted on an Argand diagram, the set of points would be illustrated by a circle of centre (0, 0) and radius 5.

The equation $|z| = 5$ is another way of describing the **locus** of points that are 5 units from the origin. In general, $|z| = r$ represents the locus of all the points that are r units from the origin.

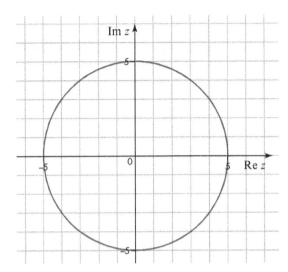

How does the locus change if the equation is written as $|z - 4| = 6$ instead? Firstly, the equation is satisfied by the real numbers 10 and -2 and the complex numbers $4 \pm 6i$.

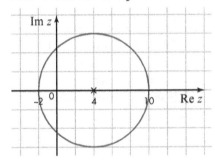

Again, you have a set of complex numbers that are equidistant, in this case 6 units, from the same point. Hence this is also a circle, but this time with a centre of $(4, 0)$ and a radius of 6.

In order to analyse the relationship between the equation and the Argand diagram, it is helpful to rewrite z as $x + iy$ and investigate the Cartesian equation.

For $|z| = 5$, you have $|x + iy| = 5$.

The modulus of $x + iy = \sqrt{x^2 + y^2}$. Hence $\sqrt{x^2 + y^2} = 5$.

Square both sides.

$x^2 + y^2 = 5^2$

which is the Cartesian equation with a circle with centre $(0, 0)$ and radius 5.

Similarly, for $|z - 4| = 6$,

$|x + iy - 4| = 6$

Group the real and imaginary terms.

$|(x - 4) + iy| = 6$

PURE MATHEMATICS 3

Hence

$$\sqrt{(x-4)^2 + y^2} = 6$$
$$(x-4)^2 + y^2 = 6^2$$

which is the Cartesian equation of a circle with centre (4, 0) and radius 6.

In general, $|z-(a+bi)| = r$ is a circle with centre (a, b) and radius r.

Example 10

Find the locus represented by the equation $|z-2+8i| = 3$.

Solution

Write $z = x + iy$

$$|x + iy - 2 + 8i| = 3$$

Group the real and imaginary terms.

$$|(x-2)+i(y+8)| = 3$$

Use Pythagoras' theorem to find the modulus.

$$\sqrt{(x-2)^2 + (y+8)^2} = 3$$
$$(x-2)^2 + (y+8)^2 = 3^2$$

This is a circle with centre (2, −8) and radius 3.

Writing $|z-2+8i| = 3$ in the form $|z-(a+bi)| = r$, it becomes $|z-(2-8i)| = 3$, which confirms that it is a circle with centre (2, −8) and radius 3.

Perpendicular bisectors

Now consider the equation $|z-3| = |z-2i|$.

Analyse the equation in a similar way to the circle.

$$|x+iy-3| = |x+iy-2i|$$

$$|(x-3)+iy| = |x+i(y-2)|$$

$$\sqrt{(x-3)^2 + y^2} = \sqrt{x^2 + (y-2)^2}$$
$$(x-3)^2 + y^2 = x^2 + (y-2)^2$$

Expand the brackets.

$$x^2 - 6x + 9 + y^2 = x^2 + y^2 - 4y + 4$$

Subtract x^2 and y^2 from both sides.

$$-6x + 9 = -4y + 4$$
$$0 = 6x - 4y - 5$$

This equation simplifies to a straight line in Cartesian coordinates.

Thinking about the original equation, it represents two sets of complex numbers with the same modulus. In fact, more specifically, it represents all the complex numbers that are equidistant from 3 on the real axis and 2i on the imaginary axis. In other words, it is the perpendicular bisector of (3, 0) and (0, 2).

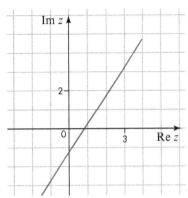

The gradient of the line segment between (3, 0) and (0, 2) is $-\frac{2}{3}$.

The gradient of the perpendicular to the line segment is $\frac{3}{2}$.

The midpoint of (3, 0) and (0, 2) is $\left(\frac{3}{2}, 1\right)$.

The equation of the perpendicular bisector is given by

$y - 1 = \frac{3}{2}\left(x - \frac{3}{2}\right)$.

Multiply through by 2.

$$2y - 2 = 3\left(x - \frac{3}{2}\right)$$

$$= 3x - \frac{9}{2}$$

Multiply through by 2 again.

$4y - 4 = 6x - 9$

$0 = 6x - 4y - 5$

This is the same equation as was produced by analysing the moduli of the complex numbers.

In general, $|z - (a + b\mathrm{i})| = |z - (c + d\mathrm{i})|$ is the perpendicular bisector of the points (a, b) and (c, d).

> **KEY INFORMATION**
> $|z - (a + b\mathrm{i})| = |z - (c + d\mathrm{i})|$ is the perpendicular bisector of the points (a, b) and (c, d).

Half lines

Consider the locus of $\arg z = 0.4$.

All complex numbers that satisfy this equation have an argument of 0.4.

Hence all these complex numbers lie on the straight line that has an argument of 0.4.

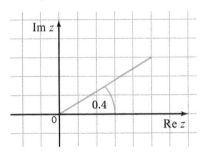

PURE MATHEMATICS 3

Note that the line starts at the origin. If the line were continued beyond the origin, these complex numbers would have a negative argument of $-(\pi - 0.4)$. As a result, this locus is called a **half line**, because although it is still infinitely long it has a starting point.

$\arg(z - 2) = \frac{2\pi}{3}$ is a half line starting at (2, 0) with an argument of $\frac{2\pi}{3}$.

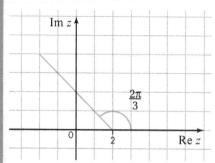

In general, $\arg(z - (a + b\mathrm{i})) = c$ is a half line starting at (a, b) with an argument of c.

Inequalities

Consider these three inequalities:

$|z| > 4$

$|z - 7| < |z + 3\mathrm{i}|$

$\frac{\pi}{6} < \arg z < \frac{\pi}{2}$.

Each of these inequalities splits the Argand diagram into two regions: one where the inequality is true and one where it is false.

$|z| > 4$ is true for any complex number with a modulus greater than 4. These points all lie outside the circumference of the circle with centre (0, 0) and a radius of 4.

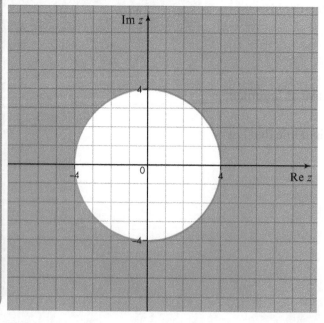

$\left|z-7\right|=\left|z+3\mathrm{i}\right|$ is a perpendicular bisector. Therefore

$\left|z-7\right|<\left|z+3\mathrm{i}\right|$ is true on one side of the line, for complex numbers closer to $(7, 0)$ than $(0, -3)$, and false on the other side.

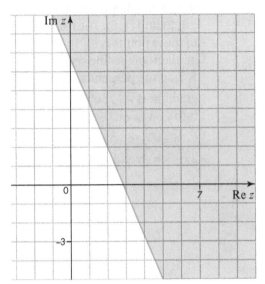

$\dfrac{\pi}{6} < \arg z < \dfrac{\pi}{2}$ is true for all complex numbers between the half

lines $\arg z = \dfrac{\pi}{6}$ and $\arg z = \dfrac{\pi}{2}$.

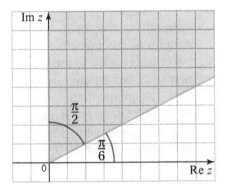

PURE MATHEMATICS 3

Example 11

a Find the locus of points for which $|z+1-2\mathrm{i}| = |z-1-6\mathrm{i}|$.

b Illustrate on an Argand diagram, the locus of points for which $|z+1-2\mathrm{i}| > |z-1-6\mathrm{i}|$.

Solution

a $|z+1-2\mathrm{i}| = |z-1-6\mathrm{i}|$

$|x+\mathrm{i}y+1-2\mathrm{i}| = |x+\mathrm{i}y-1-6\mathrm{i}|$

Group the real and imaginary terms.

$|(x+1)+\mathrm{i}(y-2)| = |(x-1)+\mathrm{i}(y-6)|$

Take the modulus of each side.

$$\sqrt{(x+1)^2+(y-2)^2} = \sqrt{(x-1)^2+(y-6)^2}$$

$$(x+1)^2+(y-2)^2 = (x-1)^2+(y-6)^2$$

$$x^2+2x+1+y^2-4y+4 = x^2-2x+1+y^2-12y+36$$

$$2x+1-4y+4 = -2x+1-12y+36$$

$$4x+8y = 32$$

$$x+2y = 8$$

This is the perpendicular bisector of the points $(-1, 2)$ and $(1, 6)$.

b Plot the points $-1+2\mathrm{i}$ and $1+6\mathrm{i}$ on the Argand diagram and the line $x+2y=8$.

$|z+1-2\mathrm{i}| > |z-1-6\mathrm{i}|$ for the complex numbers that are closer to $(1, 6)$ than $(-1, 2)$.

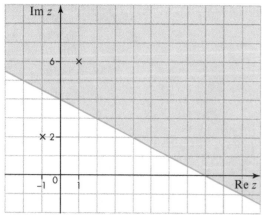

By rewriting the equation as $|z-(-1+2\mathrm{i})| = |z-(1+6\mathrm{i})|$, you can see that the locus is the perpendicular bisector of $(-1, 2)$ and $(1, 6)$. The midpoint of these points is $(0, 4)$. The gradient of the line segment is 2, so the perpendicular gradient is $\frac{-1}{2}$. Substituting into $y-y_1 = m(x-x_1)$, you get $y-4 = \frac{-1}{2}(x-0)$, from which $2y-8 = -x$ and $x+2y=8$.

Alternatively, choose a complex number (such as $0+0\mathrm{i}$) and check it in the inequality. For example, is $|0+1-2\mathrm{i}|$ greater than $|0-1-6\mathrm{i}|$? $\sqrt{5} > \sqrt{37}$ is not true, so the inequality is true on the other side of the line.

Exercise 9.6A

1 Show that the equation $|z-2+8i|=13$ describes a circle and find the centre and radius of the circle.

C 2 Show that the equation $|z-3|=|z-2-i|$ describes a straight line and find the Cartesian equation of the line.

MM 3 Describe the locus given by each equation. For each question, you should use $z = x + iy$ to deduce the Cartesian equation and plot the locus on an Argand diagram.

    a $|z| = 10$ **b** $|z - 9| = 4$

    c $|z + 2i| = 8$ **d** $|z - (5 - 6i)| = 2$

    e $|z + 4 - 3i| = 7$

MM 4 Describe the locus given by each equation. For each question, you should use $z = x + iy$ to deduce the Cartesian equation and plot the locus on an Argand diagram.

    a $|z - 5| = |z - 3i|$ **b** $|z + i| = |z - 4|$

    c $|z + 3i| = |z - 7i|$ **d** $|z - 2 - 4i| = |z + 5 - 2i|$

    e $|z + 5 + 2i| = |z - 7 - 6i|$

MM 5 Illustrate the following on an Argand diagram.

    a $\arg z = 1$ **b** $\arg(z + 3) = \dfrac{\pi}{4}$

    c $\arg(z - 1 + 2i) = \dfrac{7\pi}{10}$ **d** $\arg(z + i) = -3$

MM 6 Illustrate the region defined by each of the following.

    a $|z - 2| < |z + 2|$ **b** $|z| < 8$ **c** $0 < \arg z < 2$

PS 7 **a** Plot on the same Argand diagram the loci of $|z - 3 - 2i| = 5$ and $|z - 6i| = |z - 7 + i|$.

    b Find the complex numbers that satisfy both $|z - 3 - 2i| = 5$ and $|z - 6i| = |z - 7 + i|$.

PS 8 Find the complex number that satisfies both $|z - 3i| = 13$ and $\arg(z - 4) = \dfrac{\pi}{4}$.

PS 9 **a** Find the complex number that satisfies each of the loci $|z - 3| = |z + 2i|$ and $|z + 3 - i| = |z - 1 + 5i|$.

    b Illustrate the region that satisfies both of the inequalities $|z - 3| > |z + 2i|$ and $|z + 3 - i| < |z - 1 + 5i|$.

10 **a** Sketch the locus of each of the following.

      i $\arg(z + 2 - 5i) = \dfrac{\pi}{4}$ **ii** $\arg(z + 2 - 5i) = -\dfrac{\pi}{2}$ **iii** $|z + 2 - 5i| = \sqrt{29}$

    b Hence illustrate the region that satisfies both $-\dfrac{\pi}{2} < \arg(z + 2 - 5i) < \dfrac{\pi}{4}$ and $|z + 2 - 5i| > \sqrt{29}$.

11 a Illustrate the region defined by $|z + a| < 2a$.

PS b Illustrate the region $|z - ai| > |z + a(2 + i)|$.

c Find, in terms of a, the complex numbers that satisfy both $|z + a| = 2a$ and $|z - ai| = |z + a(2 + i)|$.

12 z is given by $5 + 2i$.

C a Show that $\dfrac{z}{z^*} = \dfrac{1}{29}(21 + 20i)$.

b By considering the arguments of z, z^* and $\dfrac{z}{z^*}$, show that $2\tan^{-1}\left(\dfrac{2}{5}\right) = \tan^{-1}\left(\dfrac{20}{21}\right)$.

c On an Argand diagram, the points O, A, B and C represent the origin, z, $z + z^*$ and z^*, respectively. What type of quadrilateral is $OABC$?

C 13 The complex number w is given by $6 + i$.

a Find w^3.

b Hence show that $3\tan^{-1}\left(\dfrac{1}{6}\right) = \tan^{-1}\left(\dfrac{107}{198}\right)$.

SUMMARY OF KEY POINTS

» $i = \sqrt{-1}$ and $i^2 = -1$.

» A complex number has a real part, Re z, and an imaginary part, Im z.

» Two complex numbers are identical if and only if their real parts are identical and their imaginary parts are identical.

» A complex number can be represented on an Argand diagram.

» The modulus can be found by using Pythagoras' theorem.

» The argument of a complex number can be found by using the tan ratio.

» To add or subtract complex numbers, add or subtract the real parts and then the imaginary parts.

» Complex numbers can be multiplied using the FOIL method. Complex numbers can be divided by multiplying the numerator and denominator by the conjugate of the denominator.

» Complex numbers have two square roots, which are also complex numbers.

» Complex roots of a polynomial equation come in conjugate pairs.

» $z = x + iy$ can be written in polar form as $r(\cos\theta + i\sin\theta)$ and exponential form as $re^{i\theta}$.

» To multiply complex numbers, multiply the moduli and add the arguments:

» $|z_1 z_2| = |z_1||z_2|$ and $\arg z_1 z_2 = \arg z_1 + \arg z_2$.

» To divide complex numbers, divide the moduli and subtract the arguments:

$$\frac{|z_1|}{|z_2|} = \frac{|z_1|}{|z_2|} \text{ and } \arg\frac{z_1}{z_2} = \arg z_1 - \arg z_2.$$

» Adding and subtracting complex numbers has the effect of translating a complex number.

» Multiplying a complex number by $a(\cos b + i\sin b)$ multiplies the modulus of the complex number by a and rotates the complex number by b in an anticlockwise direction about the origin.

» Dividing a complex number by $a(\cos b + i\sin b)$ divides the modulus of the complex number by a and rotates the complex number by b in a clockwise direction about the origin.

» $|z - (a + bi)| = r$ is a circle of centre (a, b) and radius r.

» $|z - (a + bi)| = |z - (c + di)|$ is the perpendicular bisector of the points (a, b) and (c, d).

» $\arg(z - (a + bi)) = c$ is a half line starting at (a, b) with an argument of c.

EXAM-STYLE QUESTIONS

1 The complex numbers w and z are defined as $-1 + 7i$ and $5 - 5i$, respectively.

   a State the complex conjugate of w.

   b Show that w and z have the same magnitude.

   c Write z in the form $r(\cos\theta + i\sin\theta)$, giving θ in terms of π.

   d Find wz.

   e Find $\arg(z + w)$, giving the answer in radians correct to 3 significant figures.

2 The complex number w is given by $\dfrac{14 - 31i}{3 - 2i}$.

(PS)

    a Show that w can be written as $8 - 5i$.

    b Given that w is a root of the equation $w^2 + cw + d = 0$, where c and d are real constants, find the values of c and d.

    c Illustrate the locus of $|z - w^*| < \sqrt{89}$ on an Argand diagram.

3 The complex numbers z_1 and z_2 are defined as $3 + 5i$ and $8 - 7i$, respectively.

(C) The complex number w is the product of z_1 and z_2.

    a Show that w is a root of the equation $x^2 - 118x + 3842 = 0$.

    b Find $\arg w$, giving the answer in radians correct to 2 decimal places.

    c Show that $w + 2z_2 - z_1$ is real.

4 The complex numbers z and w are given by $z = (17 - i)$ and $w = (2 + 5i)$.
Find:

    a $\arg z$, in the domain $-\pi \leqslant \theta < \pi$ **b** $|w|$, correct to 3 significant figures

    c zw **d** $\dfrac{z}{w}$.

5 The complex numbers z_1 and z_2 are defined by $z_1 = 19 - 9i$ and $z_2 = 3 + 2i$.

    a Show that the cube of z_2 is given by $-9 + 46i$.

    b Show that $|z_1 + z_2| = \sqrt{533}$.

    c Show that $\dfrac{z_1}{z_2}$ is given by $3 - 5i$.

    d Find the locus of points given by $|z - z_1| = |z - z_2|$.

    Write the answer in the form $ax + by + c = 0$, where a, b and c are integers.

6 The complex number w is such that $w = \dfrac{1}{2} + \dfrac{\sqrt{3}}{2}i$.

    a Find the modulus of w.

    b Find the argument of w in terms of π.

    c Show that $w^3 = -1$.

    d Hence, or otherwise, find the three roots of the equation $z^3 = -1$.

    e Show the three roots on an Argand diagram.

7 The complex numbers w_1 and w_2 are defined as $w_1 = -7 + 7i$ and $w_2 = 7 + 7i$.

    a Find, in the form $|z - a| = k$, the equation of the circle passing through w_1, w_2 and the origin.

    b Write w_1 in the form $r(\cos \theta + i \sin \theta)$, giving r and θ as exact values.

    c Find the value of w_1^8.

8 The point M represents the complex number $z_1 = 1 - 8i$.

The point N represents the complex number $z_2 = 4 + 7i$.

  a Show that the triangle OMN is isosceles.

  b Show that the cosine of the angle MON is $-\frac{4}{5}$.

  c Hence find area of the triangle OMN.

9 The complex number w is defined by $\dfrac{18\left(\cos\frac{2\pi}{5} + i\sin\frac{2\pi}{5}\right)}{3\left(\cos\frac{7\pi}{10} - i\sin\frac{7\pi}{10}\right)}$.

  a Write w in the form $r(\cos\theta + i\sin\theta)$, giving the argument in terms of π in the domain $-\pi < \theta < \pi$.

  b Display w on an Argand diagram.

  z is the complex number such that $wz = -12$.

  c Write z in the form $re^{i\theta}$, giving θ in terms of π in the domain $-\pi < \theta < \pi$.

10 Illustrate the region on an Argand diagram that satisfies both the inequalities
$|z - 6| > |z - 4 - 6i|$ and $|z - 5 - 3i| < 3$.

11 z is defined such that $\dfrac{26 + 29i}{z} = 6 + i$.

  a Find z in the form $x + iy$.

  b Find $\arg(iz)$, in radians correct to 3 significant figures in the domain $0 < \theta < 2\pi$.

  z is a root of the polynomial equation $5z^3 + 195z + 41 = 49z^2$.

  c Find the sum of the roots of the equation.

12 The complex number z is given by $45 - 28i$.

  a Write z in the form $re^{i\theta}$, giving θ in radians correct to 3 significant figures.

  b Find the modulus of z^3.

  c i Find the square roots of z, writing the answers in the form $x + iy$.

  ii Display the points representing the square roots of z on an Argand diagram.

13 The complex number z is defined as $(2 - i)^3$.

  a Find z in the form $x + iy$.

  You are now given that z is a root of the polynomial equation $z^3 - 2z^2 + az + 250 = 0$.

  b Find the value of a.

  c Find the real root of the equation.

  The complex number w is defined as $\dfrac{z}{4 + 3i}$.

  d Show that w can be written as $-1 - 2i$.

14 a Find the locus of all the points that satisfy the equation $|z - 8| = |z + 2i|$.

Give the answer in the form $ax + by = c$, where a, b and c are integers.

b Find the complex number that satisfies $|z - 8| = |z + 2i|$ and $\arg(z + 3 - 6i) = -\frac{\pi}{4}$.

The region R is given by $|z - 8| > |z + 2i|$ and $-\frac{\pi}{4} < \arg(z + 3 - 6i) < \frac{\pi}{4}$.

c Illustrate R on an Argand diagram.

d Find the area of R.

15 Complex numbers z_1, z_2 and z_3 are given by $z_1 = -6 + 7i$, $z_2 = 3 + 5i$ and $z_3 = 9 - 2i$.
The points P, Q and R represent the complex numbers z_1, z_2 and z_3, respectively.

a Plot the points on an Argand diagram.

b Show that $OPQR$ is a rhombus.

c Find the angle PQR, writing the answer in degrees correct to 1 decimal place.

d Find the area of $OPQR$.

16 The complex numbers z and w are given by $3(\cos 0.5 + i \sin 0.5)$ and $4(\cos 0.4 + i \sin 0.4)$, respectively.

a Write zw in the form $r(\cos \theta + i \sin \theta)$.

b Write the square roots of w in the form $r(\cos \theta + i \sin \theta)$.

c Given that $|z^m w^n| = 432$, find $\arg(z^m w^n)$

17 The points A, B, C and D represent the roots of the quartic equation
$z^4 - 6z^3 + 14z^2 - 64z + 680 = 0$.

One of the roots of the equation is $-2 + 4i$.

a Find the other three roots.

b Plot all four roots on an Argand diagram.

c Find the area of the quadrilateral $ABCD$.

d Find the equation of the circle passing through points A, B, C and D, giving the answer in the form $|z - a| = b$, where a and b are integers.

e Illustrate the region $|z - a| > b$.

18 For the complex number $z = -8 + 8i$, find:

a $|z^*|$

b $\arg z^*$, in the domain $-\pi < \theta < \pi$.

The complex number $w = a + 2i$.

c Given that $|z + w| = 26$ and $a < 0$, find the value of a.

19 The complex number z is given by $\dfrac{7+6i}{1+3i}$.

   a Write z in the form $x + iy$.

   b Find arg z.

   c Illustrate on an Argand diagram the inequality arg $(z + 2i) > \dfrac{\pi}{3}$.

   z is one of the roots of the equation $6z^3 + 11z + 68 = 22z^2$.

   d Find the other two roots of the equation.

20 **a** Expand and simplify $z = (2 + pi)^4$, giving the answer in terms of p.

   b Given that Re $z = 41$, find both possible values for p.

   c Given further that $p < 0$, find Im z.

Mathematics in life and work

In an AC circuit, the voltage across a resistor is in phase with the current. The voltage across a capacitor leads by 90° and the voltage across an inductor lags by 90°. The reactance (R), capacitive reactance (X_C) and inductive reactance (X_L) can be represented on an Argand diagram as shown below.

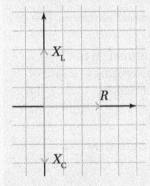

The impedance (Z) is the overall resistance and is given by the modulus of the complex number $R + i(X_L - X_C)$.

An AC circuit has a reactance of $10\,\Omega$, a capacitive reactance of $5\,\Omega$ and an inductive reactance of $29\,\Omega$.

1 Find the magnitude of the impedance.

2 Write the impedance in polar form.

SUMMARY REVIEW

Practise the key concepts and apply the skills and knowledge that you have learned in the book with these carefully selected past paper questions supplemented with exam-style questions and extension questions written by the authors.

Warm-up Questions	A Level Questions	Extension Questions
Three Cambridge A Level Mathematics past paper questions based on prerequisite skills and concepts that are relevant to the main content of this book.	Selected past paper exam questions and exam-style questions on the topics covered in this syllabus component.	Extension questions that give you the opportunity to challenge yourself and prepare you for more advanced study.

Warm-up questions

1 An oil pipeline under the sea is leaking oil and a circular patch of oil has formed on the surface of the sea. At midday the radius of the patch of oil is 50 m and is increasing at a rate of 3 metres per hour. Find the rate at which the area of the oil is increasing at midday. **[4]**

Cambridge International AS & A Level Mathematics 9709 Paper 11 Q3 Nov 2012

2 Relative to an origin O, the position vectors of three points, A, B and C, are given by

$$\overrightarrow{OA} = \mathbf{i} + 2p\mathbf{j} + q\mathbf{k}, \quad \overrightarrow{OB} = q\mathbf{j} - 2p\mathbf{k} \text{ and } \overrightarrow{OC} = -(4p^2 + q^2)\mathbf{i} + 2p\mathbf{j} + q\mathbf{k},$$

where p and q are constants.

  i Show that $\overrightarrow{OA}$ is perpendicular to $\overrightarrow{OC}$ for all non-zero values of p and q. **[2]**

  ii Find the magnitude of $\overrightarrow{CA}$ in terms of p and q. **[2]**

  iii For the case where $p = 3$ and $q = 2$, find the unit vector parallel to $\overrightarrow{BA}$. **[3]**

Cambridge International AS & A Level Mathematics 9709 Paper 11 Q6 June 2013

3 The coordinates of A are $(-3, 2)$ and the coordinates of C are $(5, 6)$. The mid-point of AC is M and the perpendicular bisector of AC cuts the x-axis at B.

  i Find the equation of MB and the coordinates of B. **[5]**

  ii Show that AB is perpendicular to BC. **[2]**

  iii Given that $ABCD$ is a square, find the coordinates of D and the length of AD. **[2]**

Cambridge International AS & A Level Mathematics 9709 Paper 11 Q9 June 2012

A Level questions:
Pure Mathematics 2

Reproduced by permission of Cambridge Assessment International Education

 i Use the trapezium rule with two intervals to estimate the value of

$$\int_0^1 \frac{1}{6 + 2e^x}\,dx,$$

giving your answer correct to 2 decimal places. [3]

ii Find $\displaystyle\int \frac{\left(e^x - 2\right)^2}{e^{2x}}\,dx.$ [4]

Cambridge International AS & A Level Mathematics 9709 Paper 21 Q6 Nov 2012

 a Find

 i $\displaystyle\int \frac{e^{2x} + 6}{e^{2x}}\,dx,$ [3]

 ii $\displaystyle\int 3\cos^2 x\,dx.$ [3]

b Use the trapezium rule with 2 intervals to estimate the value of

$$\int_1^2 \frac{6}{\ln(x + 2)}\,dx,$$

giving your answer correct to 2 decimal places. [3]

Cambridge International AS & A Level Mathematics 9709 Paper 21 Q6 Nov 2013

3 **i** Show that $\displaystyle\int_6^{16} \frac{6}{2x - 7}\,dx = \ln 125.$ [5]

ii Use the trapezium rule with four intervals to find an approximation to

$$\int_1^{17} \log_{10}x\,dx,$$

giving your answer correct to 3 significant figures. [3]

Cambridge International AS & A Level Mathematics 9709 Paper 21 Q6 Nov 2014

A Level questions: Pure Mathematics 2
and Pure Mathematics 3

Reproduced by permission of Cambridge Assessment International Education

 Use logarithms to solve the equation

$$5^{x + 3} = 7^{x-1},$$

giving the answer correct to 3 significant figures. [4]

Cambridge International AS & A Level Mathematics 9709 Paper 21 Q1 Nov 2015

5 The equation of a curve is

$$y = 6 \sin x - 2 \cos 2x.$$

Find the equation of the tangent to the curve at the point $\left(\frac{1}{6}\pi, 2\right)$. Give the answer in the form $y = mx + c$, where the values of m and c are correct to 3 significant figures. [5]

Cambridge International AS & A Level Mathematics 9709 Paper 21 Q3 June 2015

6 i Find $\int 4 \cos^2\left(\frac{1}{2}\theta\right) d\theta.$ [3]

ii Find the exact value of $\int_{-1}^{6} \frac{1}{2x+3}\, dx.$ [4]

Cambridge International AS & A Level Mathematics 9709 Paper 21 Q3 Nov 2014

7 The polynomials f(x) and g(x) are defined by

f(x) = $x^3 + ax^2 + b$ and g(x) = $x^3 + bx^2 - a$,

where a and b are constants. It is given that $(x + 2)$ is a factor of f(x). It is also given that, when g(x) is divided by $(x + 1)$, the remainder is -18.

i Find the values of a and b. [5]

ii When a and b have these values, find the greatest possible value of g(x) − f(x) as x varies. [2]

Cambridge International AS & A Level Mathematics 9709 Paper 21 Q4 June 2015

8 i By sketching a suitable pair of graphs, show that the equation

$$3 \ln x = 15 - x^3$$

has exactly one real root. [3]

ii Show by calculation that the root lies between 2.0 and 2.5. [2]

iii Use the iterative formula $x_{n+1} = \sqrt[3]{(15 - 3 \ln x_n)}$ to find the root correct to 3 decimal places. Give the result of each iteration to 5 decimal places. [3]

Cambridge International AS & A Level Mathematics 9709 Paper 21 Q4 June 2014

9 a Prove that $2 \operatorname{cosec} 2\theta \tan \theta \equiv \sec^2 \theta$. [3]

b Hence

i solve the equation $2 \operatorname{cosec} 2\theta \tan \theta = 5$ for $0 < \theta < \pi$, [3]

ii find the exact value of $\int_0^{\frac{1}{6}\pi} 2 \operatorname{cosec} 4x \tan 2x\, dx.$ [4]

Cambridge International AS & A Level Mathematics 9709 Paper 21 Q6 June 2015

10 Solve the inequality $|2x - 5| > 3|2x + 1|$. [4]

Cambridge International A Level Mathematics 9709 Paper 31 Q1 Nov 2015

11 Using the substitution $u = 3^x$, solve the equation $3^x + 3^{2x} = 3^{3x}$, giving your answer correct to 3 significant figures. [5]

Cambridge International A Level Mathematics 9709 Paper 31 Q2 Nov 2015

12 **i** Simplify $\sin 2\alpha \sec \alpha$. [2]

   ii Given that $3\cos 2\beta + 7\cos \beta = 0$, find the exact value of $\cos \beta$. [3]

Cambridge International A Level Mathematics 9709 Paper 31 Q1 June 2014

13 The parametric equations of a curve are

$$x = \frac{1}{\cos^3 t}, \; y = \tan^3 t,$$

where $0 \leqslant t < \frac{1}{2}\pi$.

   i Show that $\dfrac{\mathrm{d}y}{\mathrm{d}x} = \sin t$. [4]

   ii Hence show that the equation of the tangent to the curve at the point with parameter t is $y = x\sin t - \tan t$. [3]

Cambridge International A Level Mathematics 9709 Paper 31 Q4 Nov 2014

14 **i** It is given that $2\tan 2x + 5\tan^2 x = 0$. Denoting $\tan x$ by t, form an equation in t and hence show that either $t = 0$ or $t = \sqrt[3]{(t + 0.8)}$. [4]

   ii It is given that there is exactly one real value of t satisfying the equation $t = \sqrt[3]{(t + 0.8)}$. Verify by calculation that this value lies between 1.2 and 1.3. [2]

   iii Use the iterative formula $t_{n+1} = \sqrt[3]{(t_n + 0.8)}$ to find the value of t correct to 3 decimal places. Give the result of each iteration to 5 decimal places. [3]

   iv Using the values of t found in previous parts of the question, solve the equation

$$2\tan 2x + 5\tan^2 x = 0$$

for $-\pi \leqslant x \leqslant \pi$. [3]

Cambridge International A Level Mathematics 9709 Paper 31 Q10 June 2012

15 It is given that $f(x) = \left|\dfrac{x}{3} - 2\right|$ and $g(x) = |2x + 6|$ for $-6 < x < 6$.

   i On the same set of axes, sketch the graphs of $y = f(x)$ and $y = g(x)$, stating clearly the coordinates of any points where the graphs meet the x-axis or the y-axis.

   ii Solve the inequality $\left|\dfrac{x}{3} - 2\right| \leqslant |2x + 6|$.

16 Solve the inequality $5^x \leqslant \dfrac{12}{5^x} - 1$.

17 Find the set of values of x for which $\ln\left(x - \dfrac{3}{2}\right) \leqslant 2\ln x - 3\ln 2$.

18

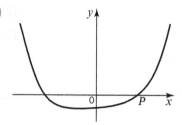

The diagram shows the curve $y = x^4 + 2x - 9$. The curve cuts the positive x-axis at the point P.

i Verify by calculation that the x-coordinate of P lies between 1.5 and 1.6. **[2]**

ii Show that the x-coordinate of P satisfies the equation

$$x = \sqrt[3]{\left(\frac{9}{x} - 2\right)}.$$ **[1]**

iii Use the iterative formula

$$x_{n+1} = \sqrt[3]{\left(\frac{9}{x_n} - 2\right)}$$

to determine the x-coordinate of P correct to 2 decimal places. Give the result of each iteration to 4 decimal places. **[3]**

Cambridge International AS & A Level Mathematics 9709 Paper 21 Q2 Nov 2013

19 i Express $5 \sin 2\theta + 2 \cos 2\theta$ in the form $R \sin (2\theta + \alpha)$, where $R > 0$ and $0° < \alpha < 90°$, giving the exact value of R and the value of α correct to 2 decimal places. **[3]**

Hence

ii solve the equation

$$5 \sin 2\theta + 2 \cos 2\theta = 4,$$

giving all solutions in the interval $0° \leqslant \theta \leqslant 360°$, **[5]**

iii determine the least value of $\dfrac{1}{(10 \sin 2\theta + 4 \cos 2\theta)^2}$ as θ varies. **[2]**

Cambridge International AS & A Level Mathematics 9709 Paper 21 Q7 June 2013

20 i Express $3 \cos \theta + \sin \theta$ in the form $R \cos (\theta - \alpha)$, where $R > 0$ and $0° < \alpha < 90°$, giving the exact value of R and the value of α correct to 2 decimal places. **[3]**

ii Hence solve the equation

$$3 \cos 2x + \sin 2x = 2,$$

giving all solutions in the interval $0° \leqslant x \leqslant 360°$. **[5]**

Cambridge International AS & A Level Mathematics 9709 Paper 21 Q7 Nov 2013

A Level questions:
Pure Mathematics 3

Reproduced by permission of Cambridge Assessment International Education

21 The straight line l_1 passes through the points $(0, 1, 5)$ and $(2, -2, 1)$. The straight line l_2 has equation $\mathbf{r} = 7\mathbf{i} + \mathbf{j} + \mathbf{k} + \mu(\mathbf{i} + 2\mathbf{j} + 5\mathbf{k})$.

  i Show that the lines l_1 and l_2 are skew. [6]

  ii Find the acute angle between the direction of the line l_2 and the direction of the x-axis. [3]

Cambridge International A Level Mathematics 9709 Paper 31 Q6 June 2015

22 i Express $\dfrac{4 + 12x + x^2}{(3 - x)(1 + 2x)^2}$ in partial fractions. [5]

  ii Hence obtain the expansion of $\dfrac{4 + 12x + x^2}{(3 - x)(1 + 2x)^2}$ in ascending powers of x, up to and including the term in x^2. [5]

Cambridge International A Level Mathematics 9709 Paper 31 Q9 June 2014

23 The variables x and θ satisfy the differential equation
$$\frac{\mathrm{d}x}{\mathrm{d}\theta} = (x + 2)\sin^2 2\theta,$$

and it is given that $x = 0$ when $\theta = 0$. Solve the differential equation and calculate the value of x when $\theta = \frac{1}{4}\pi$, giving your answer correct to 3 significant figures. [9]

Cambridge International A Level Mathematics 9709 Paper 31 Q8 Nov 2015

24 The complex number w is defined by $w = \dfrac{22 + 4\mathrm{i}}{(2 - \mathrm{i})^2}$.

  i Without using a calculator, show that $w = 2 + 4\mathrm{i}$. [3]

  ii It is given that p is a real number such that $\frac{1}{4}\pi \leqslant \arg(w + p) \leqslant \frac{3}{4}\pi$. Find the set of possible values of p. [3]

  iii The complex conjugate of w is denoted by w^*. The complex numbers w and w^* are represented in an Argand diagram by the points S and T respectively. Find, in the form $|z - a| = k$, the equation of the circle passing through S, T and the origin. [3]

Cambridge International AS & A Level Mathematics 9709 Paper 31 Q8 June 2015

25 Find the exact value of $\displaystyle\int_1^4 \frac{\ln x}{\sqrt{x}}\,\mathrm{d}x$. [5]

Cambridge International A Level Mathematics 9709 Paper 31 Q3 Nov 2013

26 i Find $\displaystyle\int \frac{2x}{x^2 + 9}\,\mathrm{d}x$.

  ii Find $\displaystyle\int \frac{2}{x^2 + 9}\,\mathrm{d}x$.

27 The variables x and y are related by the differential equation

$$\left(e^y \sin y\right)\left(x^2 + 1\right)\frac{dy}{dx} = 1.$$

Find an expression for the general solution of x in terms of y.

28 The complex number $1 + (\sqrt{2})i$ is denoted by u. The polynomial $x^4 + x^2 + 2x + 6$ is denoted by $p(x)$.

  i Showing your working, verify that u is a root of the equation $p(x) = 0$, and write down a second complex root of the equation. [4]

  ii Find the other two roots of the equation $p(x) = 0$. [6]

Cambridge International A Level Mathematics 9709 Paper 31 Q9 Nov 2012

29 The straight line l_1 passes through the points $(-8, -1, 8)$ and $(7, -1, -4)$. The straight line l_2 passes through the points $(5, -6, -7)$ and $(-1, 4, 7)$. l_1 and l_2 intersect at the point A.

  i Find the vector equations for l_1 and l_2.

  ii Find the coordinates of A.

  iii Find the acute angle between l_1 and l_2. Give your answer correct to 3 significant figures.

30 **i** Expand $\dfrac{1}{\sqrt{(1 - 4x)}}$ in ascending powers of x, up to and including the term in x^2, simplifying the coefficients. [3]

  ii Hence find the coefficient of x^2 in the expansion of $\dfrac{1 + 2x}{\sqrt{(4 - 16x)}}$. [2]

Cambridge International A Level Mathematics 9709 Paper 31 Q2 June 2012

31 When $(1 + ax)^{-2}$, where a is a positive constant, is expanded in ascending powers of x, the coefficients of x and x^3 are equal.

  i Find the exact value of a. [4]

  ii When a has this value, obtain the expansion up to and including the term in x^2, simplifying the coefficients. [3]

Cambridge International A Level Mathematics 9709 Paper 31 Q4 Nov 2012

32 Throughout this question to use of a calculator is not permitted.

  i The complex numbers u and v satisfy the equations

$$u + 2v = 2i \quad \text{and} \quad iu + v = 3.$$

Solve the equations for u and v, giving both answers in the form $x + iy$, where x and y are real. [5]

  ii On an Argand diagram, sketch the locus representing complex numbers z satisfying $|z + i| = 1$ and the locus representing complex numbers w satisfying $\arg(w - 2) = \frac{3}{4}\pi$. Find the least value of $|z - w|$ for points on these loci. [5]

Cambridge International AS & A Level Mathematics 9709 Paper 31 Q8 Nov 2013

33 **i** Show that $\int_2^4 4x \ln x \, dx = 56 \ln 2 - 12$. [5]

ii Use the substitution $u = \sin 4x$ to find the exact value of $\int_0^{\frac{1}{24}\pi} \cos^3 4x \, dx$. [5]

Cambridge International A Level Mathematics 9709 Paper 31 Q8 June 2013

Extension questions

1 $f(x) = |x^2 - 5|$ and $g(x) = |3x - 1|$

i On the same set of axes, sketch the graphs of $y = f(x)$ and $y = g(x)$, showing clearly the coordinates of any points where the graphs meet the coordinates axes.

ii Solve the inequality $f(x) < g(x)$, showing all your working.

2 **i** Show that $\dfrac{x^3}{(x+1)(x+2)} \equiv (x - 3) + \dfrac{7x + 6}{(x+1)(x+2)}$.

ii Hence express $\dfrac{x^3}{(x+1)(x+2)}$ in partial fractions.

3 Find the set of values of a for which the equation $e^{\sin x + \cos x} = a$ has a solution.

4 **i** Find the exact value of
$$\ln(e^{\pi}) + \ln(e^{2\pi}) + \ln(e^{3\pi}) + \ln(e^{4\pi}) + \cdots + \ln(e^{10\pi}).$$

ii Find an expression in terms of n for the exact value of
$$\ln(e^{\pi}) + \ln(e^{2\pi}) + \ln(e^{3\pi}) + \ln(e^{4\pi}) + \cdots + \ln(e^{n\pi}).$$

5 Find the set of values of x for which the gradient of the graph $y = \tan^{-1} x$ is greater than the gradient of the graph $y = \frac{1}{3}x^3 - 4x + \ln 5$.

6 Decide whether the following trigonometric statements are always true, sometimes true or never true.

i $\operatorname{cosec}^2 \theta + \tan^2 \theta = \sec^2 \theta + \cot^2 \theta$

ii $1 - \cos 2\theta = \cot \theta \sin 2\theta$

iii $\cos^2 2\theta = \sin^2 2\theta + \cos 4\theta$

iv $\cos^2 2\theta = \sin^2 2\theta + \sin 4\theta$

7 **i** By writing $\cos(A + B + C)$ as $\cos((A + B) + C)$ show that
$$\cos(A + B + C) \equiv \cos A \cos B \cos C - \sin A \sin B \cos C - \sin A \cos B \sin C - \cos A \sin B \sin C$$

ii Given that $\dfrac{13\pi}{12} = \left(\dfrac{\pi}{2} + \dfrac{\pi}{3} + \dfrac{\pi}{4}\right)$, find the exact value of $\cos\left(\dfrac{13\pi}{12}\right)$.

8 The variables x and y are related by the differential equation

$$\frac{dy}{dx} = \frac{1}{(x^2+1)\tan^{-1}x} \qquad 0 < x < \pi$$

When $x = 1$, $y = \ln\frac{\pi}{2}$. Find an expression for y in terms of x.

9 Given that $z = 3 + 2i$ is a root of the polynomial equation

$$z^5 - 4z^4 - 6z^3 + 72z^2 - 115z + 52 = 0$$

find the other 4 roots.

10 A locus of points within the complex plane is defined by

$$\frac{\pi}{4} \leqslant \arg(z) \leqslant \tan^{-1}\left(\frac{7}{3}\right) \qquad \text{and} \qquad |z - 3 - 5i| \leqslant 2$$

Find the minimum and maximum possible values of $|z|$.

11 A triangle has vertices A(0, 0, 0), B(1, –5, 2) and C(4, 0, –1). Find the exact area of the triangle.

12 Let $f(x) = e^{2x-1} + 2x - 6$.

i When $f(x) = 0$, identify an interval in the form $a < x < a + 1$, where a is an integer.

ii Using an iterative method with an appropriate value for x_0, solve $f(x) = 0$, giving your answer correct to 3 significant figures.

GLOSSARY

absolute value   See **modulus**.

Argand diagram   An Argand diagram is a grid with real numbers on the horizontal axis and imaginary numbers on the vertical axis.

argument   The argument of a complex number is the angle the complex number makes with the positive horizontal axis on the Argand diagram. The argument of a complex number z is written arg z. It is usually given in the range $-\pi < \theta \leqslant \pi$ radians.

base   Every statement using logarithms has an equivalent statement using indices. If $y = a^x$, then $\log_a y = x$. In both of these statements a is referred to as the base.

bearing   An angle measured clockwise from north.

binomial expansion   This is used to multiply out (expand) brackets, such as $(a + b)^n$, quickly.

If n is a positive integer then $(a + b)^n = a^n + na^{n-1}b + \frac{n(n-1)}{2}a^{n-2}b^2 + \dots + b^n$. This is a binomial expansion and has a finite number of terms.

If $a = 1$, and $b = x$, then the expansion becomes $(1 + x)^n = 1 + nx + \frac{n(n-1)}{2}x^2 + \frac{n(n-1)(n-2)}{3!}x^3 + \dots$
If n is a rational number and $|x| < 1$, then it is the general binomial expansion and gives an infinite series that will converge.

cobweb diagram   A type of diagram that shows graphically the iterations of a formula. Subsequent iterations spiral in towards the root, producing a diagram that looks like a cobweb.

collinear   Points are said to be collinear if they all lie on the same straight line.

column vector   A way of writing a vector in three dimensions, $\begin{pmatrix} a \\ b \\ c \end{pmatrix}$, where a, b, and c are distances in the directions of the Cartesian axes, x, y, and z, respectively.

complex number   A number of the form $x + iy$, where x is the real part, y is the imaginary part and i is the square root of -1.

conjugate   The conjugate of a complex number has the same real part but the opposite sign for the imaginary part. For example, $3 + 2i$ is the conjugate of $3 - 2i$.

constant of proportionality   A multiplier linking two variables.

convergent sequence   A sequence is convergent if the nth term gets closer and closer to some limit as n increases.

cosecant   The reciprocal of the sine of an angle. It is usually abbreviated to cosec or csc and so $\operatorname{cosec}\theta = \frac{1}{\sin\theta}$.

cotangent   The reciprocal of the tangent of an angle. It is usually abbreviated to cot and so $\cot\theta = \frac{1}{\tan\theta}$.

differential equation   An equation containing one or more derivatives of a function. Solving a differential equation involves integration.

divergent sequence   A sequence is divergent if values increase in magnitude without limit as n increases.

divisor   When you carry out a division, the divisor is what you divide by. For example, when $x^3 + 4x^2 + x - 6$ is divided by $(x - 1)$ the divisor is $(x - 1)$.
$F(x) \equiv Q(x) \times \text{divisor} + \text{remainder}$,
where $F(x)$ is the original polynomial and $Q(x)$ is the quotient.

double angle formula   An identity that expresses a trigonometric function of a double angle (for example $2A$) in terms of trigonometric functions of the single angle. The double angle formulae are
$\sin 2A \equiv 2\sin A\cos A$,
$\cos 2A \equiv \cos^2 A - \sin^2 A$,
$\tan 2A = \frac{2\tan A}{1 - \tan^2 A}$.

equal vectors   Two vectors are equal if they have the same magnitude and the same direction.

exponential growth or decay   Change over time represented by an exponential function. The rate of change of the function is proportional to the value of the function.

factor theorem   The factor theorem says that if $f(a) = 0$, then $(x - a)$ is a factor of $f(x)$.

foot   The shortest distance from a point P to a line L, where P is not on the line, is found by drawing a perpendicular line from P to L. The point at which the perpendicular meets the line is called the foot of the perpendicular.

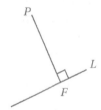

The diagram shows the foot of the perpendicular labelled F.

general solution   The general solution of a differential equation is an equation containing an arbitrary constant. It will correspond to a family of curves. For example, the general solution of the differential equation $\frac{dy}{dx} = x$ is $y = \frac{1}{2}x^2 + c$, which corresponds to a set of parabolas.

half line   The set of complex numbers of the form $\arg(z - (a + bi)) = \alpha$ is an infinitely long line starting at the point $(a + bi)$ at an angle of α to the horizontal. This is called a half line as it is of infinite length and has one end point.

i j k notation   A way of writing a vector in three dimensions, where **i**, **j** and **k** are unit vectors in the directions of the Cartesian axes x, y, and z, respectively.

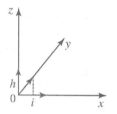

For example, $(2\mathbf{i} + 3\mathbf{j} - \mathbf{k})$ represents a vector 2 units in the x direction, 3 units in the y direction and 1 unit in the negative direction of the z-axis. In this example, 2, 3, and –1 are the coefficients of **i**, **j**, and **k**, respectively.

imaginary part   When a complex number is written in the form $x + iy$, y is the imaginary part. For a complex number z, the imaginary part is written Im z.

implicit equation   An equation which gives a relationship that the dependent and independent variables must satisfy, rather than defining the dependent variable directly (explicitly). That is, it is not written in the explicit form of $y = f(x)$. For example, a circle may have the implicit equation $x^2 + y^2 = 100$.

index   The power to which a base number is raised; in 3^4, 4 is the index and 3 is the base number.

integrand   The function that is being integrated in an integral.

integration by parts   A method used to integrate a product of two functions, u and $\frac{dv}{dx}$, using the rule

$$\int u \frac{dv}{dx}\, dx = uv - \int v \frac{du}{dx}\, dx.$$

The aim is to change a difficult integration into an easier one.

intersection   The point at which two lines (or a line and a plane) meet.

iteration   The repetition of a process or action. You can use iteration to find a sequence of approximations, each one getting closer to a root of f(x) = 0.

iterative formula   A formula that allows the next term in a sequence of approximations to be calculated from the previous term. It is usually given in the form $x_{n+1} = g(x_n)$.

locus   A set of points that satisfy a rule.

logarithm   The logarithm of a number is the power of the base that will give that number. If $a^x = b$ then $x = \log_a b$.

magnitude   The magnitude of a number is its size, for example, the magnitude of –3 is 3. The magnitude of a vector is its length. The magnitude of a complex number is the length of its representation on an Argand diagram. Pythagoras' theorem is used to find the magnitude of a vector and to find the magnitude of a complex number.

mid-point   The position vector of the mid-point of the line segment AB is $\frac{1}{2}(\overrightarrow{OA} + \overrightarrow{OB})$.

modulus   The modulus of a real number is its positive numerical value, or the magnitude of the number. It is shown by two vertical lines. For example, $|3.5| = |-3.5| = 3.5$. The modulus (magnitude) of a complex number can be found by using Pythagoras' theorem.

natural logarithm   A logarithm to base e, where e = 2.718 281…. If $e^x = a$ then $\ln a = x$.

opposite vectors   Two vectors are opposite if they have the same magnitude but have opposite directions.

origin   The point (0, 0).

parallel vectors   Two vectors are parallel if they have the same direction. If two vectors are parallel, then each can be written as a multiple of the other.

parallelogram law   If the vectors **a** and **b** are drawn in that order from a point P and then the vectors **a** and **b** are also drawn in the opposite order starting from the same point P, a parallelogram is produced. This is called the parallelogram law.

parameter   An extra variable used to express a functional relationship between two other variables, usually x and y. Parameter is also the name given to constants introduced when trying to fit a model to data.

parametric equation   An equation that involves an independent variable in which each coordinate of a point on a graph can be expressed. For example $x = 5\cos t$, $y = 5\sin t$ is a parametric equation where the variable t is called a parameter.

partial fraction   One of two or more fractions into which a more complex fraction can be split. For example $\frac{3x - 1}{x^2 - 1} \equiv \frac{1}{x - 1} + \frac{2}{x + 1}$.

The fractions on the right-hand side are called partial fractions.

particular solution   A particular solution of a differential equation is one that satisfies a particular condition. For example, it might pass through a particular point.

polar form   A complex number written in the form $r(\cos \theta + i \sin \theta)$ is said to be in polar form, where r is the magnitude of the complex number and θ is the argument of the complex number.

polynomial $P(x) = a_0 x^n + a_1 x^{n-1} + \cdots a_n$, where $a_0 \neq 0$ and n is a positive integer.

position vector The relative displacement of a point from the origin. For example, a point with coordinates (–2, 3) has the position vector (–2**i** + 3**j**).

product rule (differentiation) A rule for differentiating the product of two functions. If $y = uv$, where u and v are functions of x, then $\frac{dy}{dx} = u\frac{dv}{dx} + v\frac{du}{dx}$.

quotient A number or expression obtained after dividing a number or expression by a divisor. $F(x) \equiv Q(x) \times$ divisor + remainder, where $F(x)$ is the original polynomial and $Q(x)$ is the quotient.

quotient rule A rule for differentiating the quotient of two functions. If $y = \frac{u}{v}$, where u and v are functions of x, then $\frac{dy}{dx} = \frac{v\frac{du}{dx} - u\frac{dv}{dx}}{v^2}$.

rate of change How quickly one variable changes with respect (relative) to another. For example, if a variable y is changing as x changes, then $\frac{dy}{dx}$ gives the rate of change of y with respect to x.

real part When a complex number is written in the form $x + iy$, x is the real part. For a complex number z, the real part is written Re z.

relative displacement The displacement of one object with respect to the position of another object.

remainder A number or expression that is left over after division by a divisor. $F(x) \equiv Q(x) \times$ divisor + remainder, where $F(x)$ is the original polynomial and $Q(x)$ is the quotient.

remainder theorem The remainder theorem says that if, $f(a) \neq 0$, then $(x - a)$ is not a factor of $f(x)$ and $f(a)$ is the remainder when $f(x)$ is divided by $(x - a)$.

roots The roots of an equation are the solutions of the equation.

scalar product The scalar product, $\mathbf{a} \cdot \mathbf{b} = |\mathbf{a}||\mathbf{b}|\cos\theta$, is a formula that can be used to find the angle between two vectors (or check if they are perpendicular). To find $\mathbf{a} \cdot \mathbf{b}$, multiply the coefficients of **i** together, then the coefficients of **j** and finally the coefficients of **k**, and then find the sum of the results.

secant The reciprocal of the cosine of an angle. It is usually abbreviated to sec and so $\sec\theta = \frac{1}{\cos\theta}$.

Series expansion The representation of a function as a sum of powers of one of its variables. For example, $f(x) = a_0 + a_1 x + a_2 x^2 + a_3 x^3 + \ldots$

separation of variables A method of solving a differential equation by rewriting it so that each side of the equation can be directly integrated with respect to one of the variables. For example, the differential equation $\frac{dy}{dx} = 2xy$ can be rewritten as $\int \frac{1}{y}\,dy = \int 2x\,dx$.

skew Two straight lines that are neither parallel nor intersecting.

staircase diagram A type of diagram that shows graphically the iterations of a formula. Subsequent iterations approach the root in steps that look like a staircase.

sub-interval An interval is a set of values such as $-1 < x < 1$. A sub-interval is an interval that is contained within another interval. For example, for the interval $-1 < x < 1$, a sub-interval could be $-0.5 < x < 0$.

trapezium rule A method of estimating area under a graph by dividing it into equal width trapezia and finding the sum of the areas of the trapezia. The trapezium rule: $\int_a^b f(x)\,dx \approx \frac{1}{2}h\{y_0 + 2(y_1 + y_2 \ldots + y_{n-1}) + y_n\}$, where $h = \frac{b-a}{n}$.

triangle law Vectors can be added using the triangle law. If the vector $\overrightarrow{AB}$ (**a**) is followed by the vector $\overrightarrow{BC}$ (**b**), then the result is the vector $\overrightarrow{AC}$ (**a** + **b**).

unit vector A vector with a magnitude of 1.

vector A quantity with a size and a direction.

INDEX

$(1 + x)^n$, binomial expansion 27–30, $\dfrac{1}{x^2 + a^2}$, integration 134–5

absolute value *see* modulus

addition of complex numbers 250
  effect on the Argand diagram 264–5

algebra
  applications 1
  binomial expansion 27–30
  dividing polynomials 11–13
  factor theorem 14
  modulus of a linear function 2–10
  partial fractions 17–25
  remainder theorem 15–16

algorithms 153, 161

area under a curve, trapezium rule 122–5

Argand diagrams 244–6
  adding and subtracting complex numbers 264–5, 268
  dividing by i 267
  dividing complex numbers 267–8
  multiplying by i 266–7
  multiplying complex numbers 265–6, 268

argument of a complex number 246–8

base of a logarithm 41

binomial approximations 28

binomial expansion 27–30, 32

calculus *see* differential equations; differentiation; integration

circles
  loci 270–2
  parametric equations 105

cobweb diagrams 163–4

collinear points 185–6

column vectors 178

complex numbers 244
  addition and subtraction 250, 264–5
  applications 242, 264
  argument 246–8
  conjugate 244–5, 246–8
  division 252, 261–2, 267
  geometric effects 264–8
  key points 279
  loci 270–6
  modulus 245–8
  multiplication 250, 260–2, 265–7
  polar form 259–62
  square roots of 254

complex roots of polynomial equations 256–8

conjugate of a complex number 244, 246–8
  roots of polynomial equations 256–8

constant of proportionality 226

cosecant (cosec) 76–7

cosine (cos)
  addition and subtraction formulae 66–9
  differentiation 91–2
  double angle formula 70–1
  integration 127, 133

cotangent (cot) 76–7

cubic equations, complex roots 256–7

damped sine waves 95, 98

differential equations
  applications 224, 230, 237, 270
  construction of 226–7
  key points 239
  separation of variables 234–6
  solution by integration 229–32

differentiation
  applications 84, 98, 102
  e^x 85–6
  implicit equations 111–13
  key points 116
  $\ln x$ 88–9
  parametric equations 105–7
  product rule 94–5
  quotient rule 98–100
  $\sin x$ and $\cos x$ 91–2
  $\tan^{-1} x$ 102–3

direction of a vector 190, 192

displacement vectors 177, 218

dividing complex numbers 252, 267, 279
  effect on the Argand diagram 267–8
  in polar form 261–2

dividing polynomials 11–13

e 44–7, 62
  differentiation of e^x 85–6
  natural logarithms 49–50

ellipses
  implicit equations 111
  parametric equations 106

equal vectors 178

equilateral triangles 218

equivalence sign (⇔) 8–9

exponential growth or decay 36, 57–8, 224, 227

$\dfrac{f'(x)}{f(x)}$, integration 136–8

factor theorem 14, 32

foot of a perpendicular 215–16

fractions, partial *see* partial fractions

general solution of a differential equation 226, 229

half lines, loci 273–4

i 244, 279
dividing by 267
multiplying by 266–7
see also complex numbers
i j vector notation 177
'iff" (⇔) 8–9
implicit equations 111
differentiation 111–13
inequalities
involving logarithms 52–3
involving moduli 8–10
loci 274–6
insulation, thermal 224, 230, 237
integrand 137
integration
$\dfrac{1}{x^2+a^2}$ 134–5
applications 121, 127, 132
$\dfrac{f'(x)}{f(x)}$ 136–8
key points 149
by parts 145–7
recognising integrals 127–9
solving differential equations 229–32
by substitution 142–3
trapezium rule 122–5
trigonometric functions 127, 132–3
using partial fractions 139–40
intersecting lines 201–2
isosceles triangles 218
iteration 161–7
cobweb diagrams 163–4
divergent sequences 169–70
key points 172
staircase diagrams 166–7
variation of methods 168–9

kites 212

loci
circles 270–2
half lines 273–4
inequalities 274–6
key points 279
perpendicular bisectors 272–3
logarithmic graphs
curve type $y = ax^n$ 55–7
curve type $y = kb^x$ 57–8
logarithms (logs) 37–9
applications 36
in bases other than 10 41–2
key points 62
laws of 38
natural logarithms 49–50, 88–9
solving equation and inequalities 52–3
logistic curves 120
logistic equation 102

magnitude of a complex number 245–8
magnitude (length) of a vector 189–92, 218
modulus 2–4
of a complex number 245–8
key points 32
solving equations and inequalities 7–10
multiplying complex numbers 250, 279
effect on the Argand diagram 265–7, 268
in polar form 260–1

natural logarithms (ln) 49–50
differentiation 88–9
Newton's law of cooling 224, 237
numerical methods
applications 153, 161, 171
finding roots 153–60
iteration 161–70
key points 172

opposite vectors 178
ordinates 124

parallel lines 201
parallel vectors 178–9, 183–4
parallelogram law, vectors 183
parallelograms 212
parametric equations 105–6
differentiation 106–7
partial fractions
key points 32
with quadratic factors 24–5
with repeated factors 21–3
use in integration 139–40
without repeated factors 17–20
particular solution of a differential equation 226, 232
perpendicular bisectors, loci 272–3
perpendicular lines 207
foot of 215–16
polar form of a complex number 259–60
division 261–2
multiplication 260–2
polynomials
division 11–13
factor theorem 14
remainder theorem 15–16
population growth 102
differential equations 226–7, 231
position vectors 194–6
product rule for differentiation 94–5, 116

quadratic equations, complex roots 256–8
quadrilaterals, identification using vectors 212–15
quartic equations 256
quotient rule for differentiation 98–100, 116
quotients 12

rate of change 226

rectangles 212

relative displacement 194

remainder theorem 15–16, 32

rhombuses 212, 213–14

right-angled triangles 218

roots of equations
- complex roots of polynomials 256–8
- iterative methods 161–70
- sign change method 153–7

scalar product of vectors 205–7, 209–10

scalene triangles 218

secant (sec) 76–8
- integration 132

separation of variables 234–6

sign change method, finding roots 153–7, 172
- failure of 158–60

sine (sin)
- addition and subtraction formulae 66–9
- damped sine waves 95, 98
- differentiation 91–2
- double angle formula 70
- integration 127, 133

skew lines 201, 203

squared numbers, useful relations 8–9

squares (quadrilaterals), properties of 212, 214–15

staircase diagrams 166–7

straight lines, vector equation of 198–9, 218
- parallel, intersecting and skew lines 201–3

substitution, integration by 142–3

subtraction of complex numbers 250
- effect on the Argand diagram 264–5

tangent (tan)
- addition and subtraction formulae 68
- differentiation of tan-1 x 102–3
- double angle formula 70
- integration 138

trapezium rule 122–5

trapeziums 212

triangle law, vectors 182

triangles, types of 218

trigonometric identities 79
- addition and subtraction formulae 66–9
- double angle formulae 70–1
- key points 81

trigonometry
- applications 65, 73, 76
- differentiation of sin x and cos x 91–2
- differentiation of tan-1 x 102–3
- integration 127, 132–3, 138
- secant, cosecant and cotangent functions 76–8, 81
- $a \sin \theta + b \cos \theta$ 73–4

unit vectors 190, 218

vector equation of a straight line 198–9, 218
- parallel, intersecting and skew lines 201–3

vector geometry
- collinear points 185–6
- mid-point rule 184
- triangle law 182

vectors
- addition and subtraction 178
- angle between two vectors 205–7, 209–10
- applications 175, 191, 215
- column vectors 178
- direction of 190, 192
- equal 178
- finding the foot of a perpendicular 215–16
- **i j** notation 177
- identifying quadrilaterals 212–15
- key points 218
- magnitude 189–92
- opposite 178
- parallel 178–9, 183–4
- parallelogram law 183
- perpendicular 207
- position vectors 194–6
- scalar product 205–7, 209–10
- unit vectors 190